# THREE ANZACS FROM MALTA

# THREE ANZACS FROM MALTA

A true story of friendship, love, and loss

GIOCONDA S. SCHEMBRI

Written and published by Gioconda S. Schembri, Australia, 2016

Cover design and typesetting by Chris Mahoney (www.chrismahoney.eu)



**National Library of Australia Cataloguing-in-Publication entry:**
Creator: Schembri, Gioconda S., author.
Title: Three Anzacs from Malta: a true story of friendship, love, and loss / Gioconda S. Schembri.

ISBN: 9780994564443 (paperback)

Subjects: Bonavia, Charles.
Beck, Waldemar.
Xuereb, Anthony.
Maltese—Australia—Biography.
Immigrants—Australia—Biography.
World War, 1914-1918—Participation, Australian.

Dewey Number: 940.30922

*In memory of Charles, Waldemar and Anthony*

*Lest we forget*

*To my beloved son Thomas,*

*hoping that, in your lifetime, you will see the day*

*when the world will finally turn its swords into ploughshares.*

# *Contents*


| | | |
|---|---|---|
| Foreword by Mark Caruana | | i |
| Prologue | The Very Beginning | iii |
| Chapter I | The Early Years | 1 |
| Chapter II | Sailing Away | 19 |
| Chapter III | In Love | 57 |
| Chapter IV | The Winds of War | 68 |
| Chapter V | In the Land of the Pharaohs | 86 |
| Chapter VI | The Dardanelles Debacle | 94 |
| Chapter VII | Somewhere in Flanders | 131 |
| Chapter VIII | Waldemar's War | 151 |
| Chapter IX | His Life for his Country | 182 |
| Chapter X | The Aftermath | 206 |
| With Gratitude | | 242 |
| Bibliography | | 244 |
| Image Credits | | 250 |

# FOREWORD

As a researcher on Maltese migration to Australia, I recommend the reading of Gioconda Schembri's book on three Maltese Anzacs who fought side-by-side with Australian troops. Why do I recommend this book?

For a number of reasons, but most importantly because she has skillfully combined documentary research with family memories by their descendants. She brings out the human dimension of history that could easily be overlooked; she does this with a strong love in the lives of three Maltese migrants who migrated to Western Australia between 1911-1912. These three Anzacs were exceptional in that all three had a university background, something unique in Maltese migration settlement.

Gioconda has a legal and consular background and is well-respected among the Maltese community in Australia, particularly in the State of Victoria. She comes from the quiet village of Mqabba, a village that also lost one of its sons, Andrew Camilleri, who was killed in action in the Western Front.

This book is the latest in a growing number of books by Maltese-Australians, living outside Malta, and further enhances the strong bond between Australians and Maltese. It is a historical biography based on facts and family recollections. It makes all Maltese proud of their contribution to their land of adoption.

Maltese by birth or descent, as well as Australians, will find delight in learning more about these three brave Maltese-Australian Anzacs. The book makes pleasant reading and showcases but three Anzacs in their contribution to the war effort.

A total of 48 Maltese joined the Australian Imperial Forces, with five more who joined the New Zealand Expeditionary Forces, and of whom, seven paid the ultimate sacrifice. Charles Bonavia, Waldemar Beck and Anthony Xuereb represent three of the finest men in Maltese-Australian military history. Lest we forget!

Mark Caruana
*Migration researcher and family historian*

# PROLOGUE

# *The Very Beginning*

**Photographic studio, Malta, 1919**

A young man in uniform stares straight at the camera. His calm demeanour belies the horrors he has witnessed in the previous four years. His name is Anthony Xuereb, known as Tony among friends.

**Boisguillaume Communal Cemetery extension, Seine-Maritime, France, 1934**

A man wearing a long overcoat and hat in hand, stands head bowed in contemplative silence, gazing at the name etched into the cold marble. It reads *W. Beck.*

**Lone Pine Memorial, Gallipoli Peninsula, Turkey, November 2000 – Author's visit**

I separate from the throng of fellow travellers that are wandering around, snapping photos and talking in hushed tones. I close my eyes and I can distinctly hear the shrill sound of shells whizzing by, the clambering of heavy feet over the top of the trenches, and the anguished crying of the wounded and dying. As nimble as a cat, I quickly make my way to the memorial and scan the multitude of names inscribed. I immediately find what I am looking for, and feel as if my heart has stopped beating. My index finger is shaking as I trace the name *Bonavia C.E.*

Who were these three men and where did they come from? What was their story and how were they connected?

This book aims to dig up the answers to these and many more questions. It tells the story of these young Maltese men - Charles Bonavia, Waldemar Beck and Anthony Xuereb - who lived during one of the most turbulent eras in history, a time when the world witnessed great progress and prosperity on one hand, and unprecedented atrocities and tragedies on the other.

The First World War saw the biggest number of countries in conflict. The war, that was supposed to end all wars, changed the face of the world forever and laid the foundation for yet another global conflict just 21 years later, which turned out to be even deadlier than the first. Between 1914 and 1918, a staggering 16.5 million people were killed, including 6.5 million civilians, and 20 million were wounded or became disabled. When reading these statistics, it is easy to fall into the trap of just thinking in terms of numbers and figures, and forget that every number represents a person. Behind these numbers were real people: fathers, sons, brothers, uncles, husbands, fiancés and boyfriends. Like any young man or woman of today, they hoped and dreamed, laughed and cried, lived and loved.

This tumultuous period of world history has always interested me, ever since I had bought a poppy from an elderly English lady while waiting for some friends on a street corner in Valletta, the capital city of Malta, some twenty odd years ago. When I asked her about the connection between the scarlet flower and the war, with great emotion she described the fields in Flanders dotted with poppies, while all around, death and carnage reigned supreme.

While Malta's pivotal role at the epicentre of the Second World War is well-documented and the subject of many books and studies, I knew little about its contribution during the 1914-1918 conflict. Furthermore, since both my parents remember WWII vividly, that era somehow always felt close and familiar to me. My mother recalls the very special day when as a nine-year old, she lined up with her classmates to have a look at the George Cross awarded to Malta and its people for their resilience and bravery. My bedtime stories also included accounts of scampering into musty damp shelters at the sound of the piercing siren that announced yet another dreaded air raid. However, by contrast, the First World War seemed always shrouded by the mist of a distant past. All World War I veterans have passed on and with them a different era has come to an end. Firsthand accounts from survivors are now impossible to come by and this aura of mystery continued to add to my interest and fuelled a quest to know more about it and the people who lived through it.

One day in 2000, at the Malta High Commission in Canberra where I was working, I came across a book called *'Gallipoli: the Malta connection'* by John Mizzi, which details Malta's role during the First World War. I was surprised to find out that among the thousands of young Australians who enlisted in that war, a number of them may have stood out from their somewhat taller and fairer Anglo-Saxon mates. These were the Maltese Anzacs, who had started to call Australia home after an arduous and long sea journey, half way around the world from a little Mediterranean island called Malta. These young Maltese men had taken active part in the historic battles of Gallipoli, in Flanders and Jutland, some of whom ended up paying the ultimate sacrifice.

Among these, three particular names leapt out at me from the pages - Charles Bonavia, Waldemar Beck and Anthony Xuereb - whose story really captured my heart. These three young friends were architecture graduates who had left their families and their small homeland behind, hoping for a better life in faraway Australia. I somehow identified myself with them. As I had done at a young age, after graduating from the University of Malta, they packed their bags and left everything that was familiar behind for larger, greener pastures. When the war broke out, all three of them answered the call for volunteers to enlist and fight for King and Country. Unknowingly, from their little corner of the world, like thousands of other young men, they were destined to play an important role in this dramatic unfolding of global events.

I was really intrigued and decided to find out more about Charles, Waldemar and Anthony. I wanted to know everything about them, what were they like, what dreams they had, what made them tick? I woke up thinking about them and fell asleep doing the same!

*1.Charles Bonavia, 2. Waldemar Beck and 3. Anthony Xuereb*

A trip to Gallipoli in November 2000 continued to pique my interest. When I came across Charles's name inscribed on the monument at Lone Pine, a seed was planted in my mind: the life story of Charles and his two brave young mates had to be told.

The dawn service at the Australian War Memorial in Canberra the following Anzac Day in 2001 held a more special significance for me. As I raised my candle in the cold misty air and joined the multitudes in praying for the dead soldiers, I murmured a silent prayer for Charles, Waldemar and Anthony. At that moment, it was almost as if I felt them nudging me to tell their story to the world. I prayed to them to give me some signs that this indeed was what they wanted me to do.

The signs came easily and in quick succession. John Mizzi's book was my point of departure. I also consulted numerous other sources about the First World War, which are listed in the bibliography section at the end of the book. A number of researchers in the field were also very helpful in pointing me in the right direction. This led me to obtain the service records of Charles, Waldemar and Anthony from the Australian National Archives. These were a mine of information. Like a detective, I sifted through the records and obtained some very significant leads. I was well on my way on this amazing journey of discovery!

While still in Australia, I contacted Waldemar's family living in Malta. Way before Google and social networking sites had become fashionable, I decided to have a look in the trusted good old Malta telephone directory. I wrote to the only Waldemar Beck listed, basing all my hope on the fact that a combination of such a name and surname are hard to come by in Malta. Surely he had to be a relative! The gamble worked perfectly. Soon after, I received a very touching email from his daughter Mariella, confirming that her great grandfather George was in fact Waldemar's only brother.

I also managed to trace the niece of Waldemar's Australian fiancée, Barbara, through one of her distant cousins, again via my old trusted friend, the telephone directory! Luckily, as fate would have it, she only lived two hours away from Canberra, where I was living at the time.

Likewise, Charles's surviving family was also just as easy to contact. It was as if all this was being orchestrated and facilitated by someone from above! Everyone was more than generous with their time and happy to share photos, war letters and stories passed down through the generations. They were all very touched by the interest in the life of their beloved relatives, about whose bravery they had heard so much.

I still recall my very first meeting with Waldemar Beck's family, upon my return to Malta in December 2001, after a three-year diplomatic posting in Australia. Admittedly, I was both nervous and excited at the prospect of meeting the living relatives of a man I had come to know and admire, albeit only from the pages of a book. During a number of encounters with the Beck family spread over a few months, I came to know Waldemar through his letters, photos and numerous family anecdotes. This was living history! In one of these meetings, I also had the opportunity to meet John Mizzi, the author of the book that had tickled my interest and fanned my appetite to know more about the Maltese Anzacs. It turned out that John also had a connection of his own with the Beck family, as it transpired that his mother was actually a cousin of Waldemar's mother, Carolina!

Anthony, however, remained an enigma. Years went by and my attempts to trace any of his family members proved fruitless. I was at a dead end, as I had absolutely no leads. I knew that, one day, all would be revealed to me. I had to wait until 2009 when I came across the name of Eligio Castaldi, who had known Anthony in Darwin. I contacted Professor Peter Castaldi, Eligio's son, who provided some very vital pieces to the puzzle with his vivid memories of Anthony visiting his father in Sydney.

Thereafter, I embarked on months and months of research, piecing all the information together, assisted at times by my mother who made a few trips to the National Library in Valletta on my behalf. The voices of Charles, Waldemar and Anthony were always present, urging me on, especially when at times my courage faltered as I felt that this mammoth task was too heavy for my little shoulders. I received other encouraging signs along the way that convinced me I was on the right path. For instance, on the 21st of January 2001, while still working at the Malta High Commission, I received a telephone query from a newly-arrived migrant from Malta. Earlier on the same day, I had read in one of historian Barry York's books that Charles, Waldemar and Anthony used to help new migrants to find jobs. I asked the person what his qualifications were: I could not believe it when he replied *'a draughtsman and surveyor'*, which had been their occupation too! The sceptics may call this a mere coincidence, but I believed this had not happened by chance and meant another nudge to *'keep on keeping on'*!

For a few years, other things took precedence and unfortunately, I had to shelve the project. Not that I ever forgot it or my three friends, who were constantly on my mind. It was not a matter of 'if' but of 'when' I would write their story. I also knew that I had to finish what I started. I had made a promise, and I was going to keep it. I owed it to myself…and to Charles, Waldemar and Anthony, as well as their families.

Initially, the idea was to write a work of fiction based on their life. But, as I delved deeper and went through all the letters, photos and records, I thought it would be a true shame if their lives were to be described as if they were just characters from a fantasy world. Although writing a novel set in this era is still on my bucket list, I felt that these men deserved to have their story told as it really happened, and not tainted by the machinations of my imaginative mind.

As I do not purport to be a migration or military history expert, I have been assisted by historians who patiently checked my work, and I have also utilised official sources and authoritative books. While I have done my utmost to ensure historical accuracy, the background information should be seen merely as a backdrop that brings these three young men to life, rather than bearing the traits of a history textbook full of facts and figures. After all, while the historical events shaped their lives, in turn they also made history.

Thus, my wish is that the book is construed more as a commemoration and a celebration of the lives of Charles, Waldemar and Anthony. I also hope that, at the same time, I have done them and their families proud. It was my privilege to get to know and admire such remarkable young men, and I hope that through this book, you will too.

# CHAPTER I

# *The Early Years*

***'The most important thing in life is not to have lived but to have deserved to live.'***
*(Entry from Waldemar's war diary)*

The apartment is located on a very busy main street in Santa Venera, a suburb in central Malta. As soon as I am welcomed into its tastefully-furnished living room, my eyes are immediately drawn to a sepia photograph of a dashingly handsome young man in uniform. A memorial plaque and scroll hang somberly on the wall and close by, a gilded frame adorns a university parchment written in Latin with the name of its recipient printed in black Gothic script: *Valdimirus Beck*. There is a feeling of peace and reverence in the room, as if I am standing at a shrine dedicated to a much-loved deceased family member, whose presence seems to linger.

The owner of the flat is the great-nephew of the man depicted in the photo. Perhaps he feels an even more special bond with him as they share the same name and surname. In an adjoining room, more photos are on display: there is Joseph or Josie, Waldemar's only nephew, being the son of his older brother George. Another image depicts Josie as a boy wearing a sailor's suit, which was such a fashion statement in the early part of the 20th century. Thousands of miles away, in the town of Mittagong, in the Southern Highlands of New South Wales in Australia, at the beautiful home of Barbara Barter née Gugeri, there exists a postcard sent by Josie to his *'dear uncle Waldemar'*. Barbara, named after her aunt, who had been Waldemar's fiancée, also owns a family portrait of George, his wife Elvira, and a six-year-old Josie, that had formed part of her aunt's collection. Moreover, another copy of this same photo exists in the collection at the apartment in Malta, on the other side of the world.

On the kitchen table, back in the Santa Venera flat, sits a cardboard box with its slightly battered lid open. Its contents are carefully and slowly taken out: more old photographs, birth and death certificates, and several well-preserved, albeit yellowing and fragile-looking letters, some still in their original envelopes. The excitement and curiosity are too much to bear. Waldemar, the great-nephew, takes a deep breath and starts talking. He is a mine of information about his relative, remembering various dates of births, deaths and marriages. It is obvious this family cherishes the late Waldemar's memory and hold him in high esteem. Through the various anecdotes, vignettes and recollections passed down through the years that Waldemar shares, his namesake is no longer just a face and a name but starts emerging as a flesh and blood young man, full of spirit and love of life. Every now and then, Waldemar pauses from his storytelling and exclaims *'Qed jerġa' jgħix'*, meaning that his great uncle is being brought back to life again. Even his daughter Mariella concurs that she also feels close to Waldemar:

*'Although I never met Waldemar I feel him near in some way; many times I keep on thinking on how he was, romantic or maybe really lonely away from home...... I feel him similar to me in many ways.'* (Email from Mariella, 21st of December 2010).

*4. Waldemar and his daughter Mariella*

Waldemar was born in the coastal town of Pietà on the wintry evening of the 11th of January 1889, at 7:30 p.m. He was the second and younger son of Lawrence and Carolina Beck née Maempel.

His older brother George had been born 11 years earlier, on the 4th of February 1878 in Floriana, the outer suburb of Valletta. The gap in years brought the two siblings closer and always made George feel protective towards his little brother.

Due to the high infant mortality rate of the time, Waldemar was baptised very quickly, when he was barely two days old, at the impressive parish church of the neighbouring town of Msida, dedicated to Saint Joseph. Father Salvatore Caruana, who had been the parish priest for more than twenty years, conducted the ceremony, while John and Maria Beck were the godparents. His baptism certificate lists a multitude of names that he was given: Waldemar Enrico Giovanni Carlo Federico Massimiliano, but he would always be called just Waldemar, Wald, or Volly for short.

*5. Lawrence Beck*

Waldemar's father Lawrence had been heavily involved in the project of building the Msida church. Although originally from Vittoriosa, one of the three cities that had always played such an important part in Malta's long and checkered history, he had moved to Msida after marrying Carolina, where he immersed himself in the local community, taking an active part in various parish activities. Being a businessman, he had numerous contacts, and with his brother-in-law, Alberto Maempel, he was a member of a committee headed by Bishop Carmelo Scicluna, set up to push forward the building of the new local parish church, which had been shackled by financial problems and disaccord on its style and size.

It seems that Waldemar's paternal grandparents, Carlo and Teresa Beck née Degiorgio, loved tradition as they named his father, as custom dictated, after the patron saint of Vittoriosa, Saint Lawrence. Lawrence, however, did not follow this tradition for his own son, as otherwise Waldemar would have been called Joseph, the revered patron saint of Msida. Rather, his parents chose a typical German first name for him in honour of his mother's German heritage. His father's family, on the other hand, originated from Ireland. In the early part of the 19th century, an Irish soldier was posted to Malta's sister island of Gozo. There the inevitable happened as he fell in love with a local girl. Their love story culminated in matrimony and after spending some time in Ireland, they went back and settled in Malta.

Waldemar's mother, Carolina, was born in Valletta, the impressive baroque capital city of Malta. Her mother, Joanna Calleja, had married Carl Heinz August Maempel in 1839 in an Anglican ceremony. Maempel became Consul of Hamburg to Malta in 1852, a position he kept until 1869. In Malta, the family always maintained the German connection and continued to observe German culture and traditions. Carolina was a great cook and she loved to make the German semolina-based recipe *'bombi tas-smied'*. Waldemar's great-nephew still treasures her well-thumbed and much-used recipe book, written in old Maltese, which has numerous margin notes in her handwriting. When Waldemar had to choose what subjects to study at the Lyceum, he

*6. Waldemar as a baby with his mother Carolina*

did not think twice in choosing the German language, as it would have come quite easily to him due to the family connection. Little did he know at the time that his knowledge would serve him in good stead years down the track, when he would use his language skills to communicate with German prisoners during the war.

One of Carolina's sisters, Vitorin, continued the link with Germany by marrying Baron Maximillian von Tucher on the 5th of June 1877, when she was 21 years old. The marriage was celebrated by both Anglican and Catholic clergy. Maximillian hailed from Leitheim in Bavaria and was the son of Baron William and Frieda von Tucher.

Von Tucher was the last German Consul in Malta before the Consulate was closed just before the commencement of hostilities. He played a very active part in the economic, social, and cultural life of Malta at the time. Amongst his various positions, he was the president of the Malta Society of Arts, Manufacturers and Commerce in 1899. At the Count Roger Band Club in Rabat, there still exists a flag, which was donated by him. As part of their household in Malta, the von Tuchers had a German cook called Bertha who spoke perfect Maltese. She used to collect folk stories from farmers and she is even mentioned in a book by Ġużè Cassar Pullicino, an eminent researcher of Maltese folklore. Conspiracy theories also abounded that Bertha was a spy for the Germans, a kind of Mata Hari in Malta!

When rumours of a looming war started to spread, and perhaps also because of Bertha's alleged covert activities, von Tucher and his Maltese wife had to leave Malta for Germany in 1913. They took with them a farmhand from Rabat, Salvu Muscat. Upon his return to Malta, Salvu used to talk all the time about his adventures overseas. He was later employed by George, when the Becks were living at 24, Inguanez Street, in Rabat. This is the house where Waldemar's great-nephew grew up and it still exists to this day. Salvu proved to be a very loyal and faithful employee. When George had a stroke in 1944, he nursed him with great devotion. He used to carry him to an armchair to sit on, this being no mean feat as George was a big man. Salvu's respect towards his employers was reciprocated as, when he retired at Ta' Sawra nursing home, Elvira would regularly send her maid, Mari, with a bowl of comforting soup for him.

Fredrick, one of Carolina and Vitorin's brothers, took over the von Tuchers' property and land in Rabat. In fact, the Beck family always referred to these properties as *'Ta' Federik'*.

Originally, these properties had been left in the care of their nephew, George, Waldemar's brother. However, when Vitorin, who was also George's godmother, happened to mention that some German cousins of her husband would expect to partake of this property, George, who was not a risk taker and disliked complications, backed out to avoid any trouble. In this respect, George and Waldemar could not have been more different. While George comes across as being serious and cautious, Waldemar seems to have had a jovial character and been a risk taker, not afraid to take the big plunge and try his luck on the opposite side of the world. Indeed, later, he would enlist in a war, which would take him back to Europe. The exchange of letters between the two brothers testify to this contrast in their personalities: while George was a worrier, Waldemar was always trying to lighten things up, in an effort to appease George's concerns.

In this respect, Waldemar had inherited his mother's cheerful character. Whenever Mary, her grandson Josie's girlfriend, visited her beloved, Carolina used to sing *'La bella figlia dell'amore'* from the opera *'Il Rigoletto'* to them. Waldemar's great-nephew fondly recalls an anecdote about a tray that Carolina owned, which also doubled as a music box. When she used to entertain guests, while serving chocolates from this tray, she would dance to the music and Waldemar joined her in this endearing mother-son dance. Since it is known that he had a sweet tooth, undoubtedly he would also have helped himself to the chocolates at the same time! Carolina

*7. Waldemar and his mother*

seems to have had a special bond with her younger son, who was born eleven years after George, when she was already in her late forties.

Waldemar's childhood, which seemed to have been idyllic, was shaken when his father Lawrence died at the young age of forty-seven on the 29th of June 1898. Being the eldest male in the house, George became the family's bread-winner, assuming the responsibilities of the '*pater familias*' when he was just twenty. He became a father figure in Waldemar's life, mentoring and advising him. Even after Waldemar moved to Australia, George continued to keep an eye on his little brother. In typical big brother fashion, George worried about Waldemar's wellbeing and financial situation, especially with him being so far away. His concern grew stronger when Waldemar went off to war. In his letters to George, Waldemar was constantly trying to put his brother's mind at rest. At the same time, he appreciated deeply George's care and concern:

'*My dearest brother,*

*God bless you for your letter of the 4th of March. If you only knew what a load it has taken off my heart. Not that you are not worrying about me. I know what you are and what love you bear me. But still George, it is very comforting to get such a letter as yours. I really thank you from the bottom of my heart.*'

'*...Dear George, thank you very much for your Christmas gift, it was just like you. But please do not send any more money. I am really rich. We get 2d a day whilst here. Then there are £5 due to me to be drawn when I get my long leave and go to London, not that there will be any need to spend it all for as I told you, I will be staying with relations of the Gugeris* (his fiancée Barbara's relatives) *so I will be all right. Then in Australia there are 4d a day being banked for me so that up to now there are more than £20 in the savings bank...I am really not in need of anything at all.*'

In the same letter, Waldemar reassured George once again:

'*I do not owe any money. There is only one little debt and I have fixed that up so I am allright dear George. But thank you all the same from the bottom of my heart.*'

He ended the letter on the same note again:

*'Good bye old chap and God bless you all. I am allright and really there is no need to worry for months and months yet.'* (Letter written from Salisbury Plains in England, 18th of March 1917)

George had married Elvira (Vira) Maistre in the chapel of Our Lady of Sorrows in Pietà on the 24th of October 1905. Vira was four years older than her husband, born on the 26th of March 1874. She had been in love with George for a long time, but her father was against the marriage as he considered the Becks of a higher social standing and was scared that George would look down on her. Mr. Maistre whisked her away to Tripoli, thinking and hoping that she would forget him in time. But this plan did not quite work out the way the concerned father had planned. A cliché it may be, but absence does make the heart grow fonder, as Vira and George discovered. When she returned to Malta, she used to watch George studying on the roof. Her love for him had not died.

A year into the marriage, on the 15th of September 1906, Vira gave birth

*8. George, Vira and Josie*

to a son called Josie. Josie was very close to Waldemar, who used to call him *'Napier'* or *'Naps'* as a term of endearment. As an adult, Josie travelled a lot. In one of the postcards sent to his father George, ironically, he writes *'All quiet on the Western Front.'*

Waldemar and his nephew were very close and exchanged a number of letters and postcards, even when Waldemar was at the Front. In a postcard addressed *'To my dear uncle Waldemar Beck'* dated the 30th of December 1916, Josie wrote:

*'Dear Uncle,*

*I wish you a very happy New Year.*

*I hope that you have had a pleasant Christmas and that you are well. I wish to see you a great deal.*

*Yours affectionately,*

*Josie Beck'*

In his letters to his brother George, Waldemar always asked after Naps. His little nephew was never far from his mind, even when he was in the midst of terrible fighting in the trenches. He also worried that Josie might not take the news that his uncle was at war too well, so he wanted to protect him from the truth. He loved Naps dearly and proudly showed off his letters and photos to his friends:

*'I am anxiously expecting Vira & Naps' letters. I do like to hear from him so much. His letters always go the round of my friends. I have separately been asked to give them his photo but I do not suppose you have got any. And how is mother? I hope doing as well as ever.'* (Letter to George, 28th of August 1916)

*'Please give my love to all. How is mother, Vira and Naps? I hope all well. Give my love to all of them. I hope you are all so well.'* (Letter dated 2nd of March 1917).

At times, Waldemar expressed his concern when news reached him that his nephew was ill:

*'In your last letter you never said anything about poor Naps. I sincerely hope that he is getting better and that expect soon to hear that all those spots have disappeared and that he is quite well again.'* (Letter to George, 28th of August 1916)

He also gave George some advice on how to help Josie come out of his shell, since he suffered from extreme shyness:

*'Lil Ġorġ għedtlu biex ma jħalliex lil Ġiusi ibati mill-mistħija. Ikun ħafna aħjar għalih.'*

*('I told George not to let Ġiusi (Josie) suffer from shyness. It is for his own good…')*

(Letter to sister-in-law Vira, 14th of May 1916)

*'I am sorry to hear about poor Naps. I am very glad to hear that he is well again. Old chap, do not forget those 2 letters I once sent you about him. See that his character gets developed along the right lines. Thank God, mine is getting changed but not until I had experienced a few nasty shocks and jars. I always pray to God that he will be spared these.'* (Letter from Salisbury Plains, England, 18th of March 1917)

*'I was very glad to hear that Naps is getting so well at school. I hope he is getting over his shyness and diffidence. You know, old chap, that it is as important as the other part of his education and I hope you will not fail to look after that part of his education. You know, really it is as important, if not more as the other part. So don't you forget it.'* (Letter dated 19th of May 1918)

Josie does not seem to have grown out of his shyness in adulthood. Mariella remembers this trait of her grandfather's character well:

*'I remember my grandfather Josie who lived with us, together with his wife Mary. I remember trying to give him some broth, he used to live practically in his own bedroom. He was a very introvert person, he used to collect so many newspapers and had hundreds of National Geographic magazines, books, stamps and coins. He used to work at the post office in Valletta and finished his career as postmaster general. My Nanna Mary was quite a positive person, I remember her very well. She was one of the few women that had a driving licence and she had her own car, while my grandfather did not drive.'* (Email from Mariella, 19th of October 2015)

Vira and George had three children in all but only Josie survived. The other two children both died when they were still under the age of one year. Vira was a very religious woman and she was particularly devoted to the *'Madonna taċ-Ċintura'* (*'Our Lady of the Garter'*). As a sign of devotion, she used to wear a belt around her waist, even up to the time when she went into labour. The midwife, who she used to call *'Kummari'*, chided her and ordered her to remove the belt as the doctor who was about to visit her, an Englishman and a non-believer, would surely be upset with her. Reluctantly, Vira removed it and when later, unfortunately, she lost the baby, she always blamed the midwife and the heathen doctor. She was very scrupulous and when she used to take a shower or bath, she used to wear a robe and wash herself through it. Four years into the marriage, Vira and George stopped sharing the marital bed and started sleeping in separate rooms. When George had his stroke, his wife lovingly took good care of him until his last breath.

When her brother's wife passed away, Vira and George took care of her daughter Jo until she was 16 years old. Their son Josie would marry his first cousin Mary, who was Jo's sister, in 1936.

Jo had many anecdotes about the Becks and she remembers the houses in Sliema very well. First, the Becks used to live in 91, Rudolph Street, which does not belong to the family anymore. Then, they moved to number 10 on the same street. This is where Waldemar used to send his letters to George from Australia and later, from the Front. Jo recalls that, as a little girl, she used to wait in anticipation at the window of the Rudolph Street house for Herman Maempel, Carolina's brother, to pass by as he always handed her sweets.

Waldemar enjoyed a very close relationship with Vira. When in Australia, he sometimes wrote directly to her. One such letter still exists. It is written in old Maltese as Vira could not speak English. Once, when an English friend of the family came to visit, the only words Vira spoke to him were what she still remembered from her school text books: *'I am very glad to see you in good health'*!

*'Din hija it-tieni ittra li jien ktibt bil-Malti minn kemm ilni hawn. Nispera li ma tkunx l-aħħar. Infatti meta ma jkollokx x'tagħmel dejjem tista' tibgħatli xi żewġ rigi.'*

*('This is the second letter I have written in Maltese since I arrived. I hope it will not be the last. When you are not busy, do drop me a few lines.')* (Letter dated 14th of May 1916)

Waldemar and Vira also bonded over their love for all things sweet. Although a diabetic, Vira kept eating cakes and sweets, which she used to hide, up to her last days. Waldemar's great-nephew recalls his grandmother talking lovingly about her brother-in-law, with particular reference to his cheerful character.

George worked at the Treasury and finished his career as Treasurer. He was a highly respected public servant and, in fact, when the Governor and his deputy were both overseas, he used to be the next in line to represent them, especially for signing purposes. He initially was a teacher, earning four shillings a day. With his meagre wage, he managed to pay Waldemar's university fees, even though, by that time, he was married and had started a family of his own.

The Melitensia collection archived at the University of Malta library includes the roll of students and shows Waldemar's attendance at the Lyceum (*'Liċeo'*) from 1900 until his graduation from the Royal University of Malta in 1911. The names of Charles Bonavia and Anthony Xuereb also appear in this list so one can assume that it was at school that these three young men, whose destinies would become so closely intertwined, met for the first time. In 1906, the three of them commenced their studies to obtain an engineering and architecture degree. Another student who attended the Lyceum at the same time was the future Prime Minister of Malta, Paul Boffa.

In the lyceum and university records, Waldemar's older brother, George, was listed as the next of kin, giving his address as 104, Molo Pieta`, Msida. His profession is listed as either clerk or civil servant. At the time, Waldemar also lived in Msida at 16, Strada Nuova.

Anthony was born on the 6th of September 1888 in Attard. Details about Anthony's father show that Federico Xuereb was a medical doctor. Between 1882 and 1900, Dr. Xuereb was the medical superintendent of the Mental Hospital in Attard and Anthony was actually born on the premises. The school records give various addresses for Anthony over the years, including 20, Strada San Marco and 70, Strada Levante in Valletta and 11, Victoria Avenue in Sliema.

*9. Charles as a young boy*

Charles was the oldest of the three friends, born in Sliema on the 25th of September 1887. He lived at 331, Strada San Paolo in Valletta. His brother, Thomas, who was two years older, also attended the Lyceum at the same time. Their father Emanuel, named as next of kin, was the registrar at the Law Courts, and their grandfather, Thomas Emanuel Bonavia, had risen to the rank of Colonel in the Royal Malta Fencible Artillery. Another of Charles's ancestors, Matteo Bonavia, was the clerk of works under the Order of St. John. He was a strong opponent of the French occupation of Malta during the last two years of the 18th century and for this, he was imprisoned at Tigné, from where he escaped and joined the British blockading force. The Maltese requested the assistance of the British and he was an interpreter accompanying Canon Francesco Caruana, one of the leaders of the Maltese insurgents, on board Nelson's flagship off Malta. More than a century later, his descendant Charles also used his language skills as an interpreter with the Australian Imperial Force (A.I.F). After the French were defeated, Matteo was attached to the Royal Engineers Department and appointed adjutant to the company of Maltese sappers and miners until he retired in 1814. He was instrumental in ending Maltese prejudice to military life, and eventually three companies of military artificers were raised to serve in Spain, Italy and Corfu.

*10. Carefree days*

An interesting photo, forming part of the Beck family collection, shows a group of young men, which included Charles, Waldemar and Anthony, pulling faces for the camera. They could even have been a bit tipsy as one of them was holding what looked like an empty bottle of wine. Perhaps they were celebrating the end of their school years. They all look carefree and enjoying their life to the full! None of them would have believed that, in a few years' time, the so-called civilised world would find itself in the grip of a bloody conflict and that they would play a role in the outcome. This photo was kindly shared by the Becks with Noel Zarb Cousin, a relative of Charles. Noel could hardly contain his excitement when he recognised Charles, as if he had found another piece of the puzzle relating to his great-uncle's life.

*11. Waldemar and Anthony*

Noel's grandmother Laura was one of Charles's sisters. She married Vincent Scicluna and her only child, Sonia, is Noel's mother. Charles was

*12. The Bonavia family - Back row: Charles's siblings Marie, Laura, Evelyn, Thomas, Melita. Front row: Charles's father, mother and sister Lily*

very close to Laura and, in fact, he used to correspond with her often from Australia. Noel and his mother's cousin, Alfred Pisani, shared a formal family photograph showing Charles's parents sitting down surrounded by their adult children. In the photo, besides Laura, there is also Thomas or Tommy, Charles's older brother, who would later migrate to the United States. One can also see Charles's other sisters: Marie (Alfred's mother), Evelyn, Melita, and:

'*The youngest one sitting down is Auntie Lily, she never got married and I remember she used to love us very much and cats too.*' (Email from Noel, 14th of March 2003)

The three young men studied a variety of subjects ranging from typewriting, to phonography, bookkeeping and physiography. As mentioned earlier, in line with his family heritage, Waldemar studied German and received prizes for proficiency in the language. One of the award certificates, dated July 1910 and signed by E. Magro in his capacity as Director of Public Instruction, still exists in the Beck family collection. The university records for the engineering and architecture course show that Waldemar kept up his German studies as an optional subject, even when pursuing his degree course.

*13. Lea Harding, the daughter of Charles's sister, Evelyn Stilon Depiro (left) and Alfred Pisani, the son of Charles's sister, Marie Pisani (right)*

These records also give information about the subjects and exams the three friends sat for, and the results they obtained. Subjects covered included algebra, trigonometry, conics, geometry, mercuration, stone cutting, land surveying, physics, mechanics, valuation, architecture and perspective, and geometrical

*14. Waldemar's school certificates*

and ornamental drawing. For their second year examinations, according to the results published on the 19th of August 1907, Waldemar obtained 380 marks, Charles scored 372, while Anthony secured 415 out of a total of 500 marks spread over five papers.

After the publication of the results for the written part, which was made in the presence of the students, they had to also sit for oral exams. In addition, their examinations contained a practical aspect. One such exam took place on the 17th of July 1909 at the Ħamrun railway station.

In July 1910, the Board of Examiners assigned by lot the area for which the students had to draw plans. Waldemar was assigned to Flores College and the first portion of Bulebel Minore. The 16th portion of the same area fell onto Charles together with the Sliema School, while Anthony's task was to draw the plans for the Żurrieq School and the 17th portion of Bulebel Minore. Besides the plans, the assignment also included the surveying of the area and the keeping of field books. All this had to be completed within twelve calendar

*15. Photo dedicated to Waldemar from his friend A. Griscti*

*16. Waldemar on his graduation*

months. The students were advised to ask the board to examine the work while in progress before the final draft of the plan was made.

In fact, over a number of months, the Board carried a number of on-site inspections to check the plans and records kept by the three candidates. After its conclusions that all the work carried out was correct, in July 1911, the three students sat for their final theoretical and practical examination and were approved, with Charles having to go through a whole gruelling hour of questions by the panel.

All the long hours of study, hard work, toil and sacrifice had finally paid off. Full of hope and aspiration, and with the parchment of their coveted architecture degree in hand, Charles, Waldemar and Anthony were looking forward to begin writing the next chapter of their life, without knowing exactly what their future held in store, and where it would take them.

# CHAPTER II

# *Sailing Away*

***'I think if some of your young lads and young men came out here, and saw what a fine country it is, they would be able to tell their friends all about it.'***

*(Dr. Richard Arthur, member of the NSW Parliament and President of the British Immigration League)*

On the 13th of August 1911, the British Star *Nyanza* sailed from Malta to Port Said. Charles Bonavia, whose final destination was Fremantle in Western Australia, was among the passengers. As the harbour lights dimmed, he would have wondered when, or even if, he would ever set eyes on his beloved Malta again.

Records held at the National Archives in Canberra show that the ship had arrived in Malta from London, and that one of its stops was Calcutta in India. At the time, Malta had no direct routes to other continents and therefore, those who wanted to travel to destinations outside Europe had to travel from Malta to Naples, Genoa, Marseilles, or Le Havre, from where they had to catch another liner.

Although accommodated in first class, Charles had to join the rest of the travellers at meal times. In his first letter to his family, written on the 16th of August when he had just been three days at sea, he mentioned four Maltese passengers who were on their way to Port Said. According to Charles, their table manners left much to be desired, to the great enjoyment of the waiters. His travel companions, Johnny Cassar and his cousin, could not take it anymore and they left the table in disgust and went out on deck.

Charles changed ships in Colombo where he embarked on the Peninsula and Orient (P&O) Steam Navigation Company ship, *China*. In another letter he wrote on the way to the Ceylonese capital, he told his father that he was nicknamed 'count' due to his foreign name and because he had *'the ways of the Count of Monte Cristo'*, referring to his elegant attire. His penchant for stylish and fashionable clothes would also get noticed when he settled in Perth, where he was once again the centre of attention because of his straw hat and light-coloured socks. Apparently, these two fashion trends had not yet reached Australian shores!

The ship manifest listed Charles as a Maltese national whose *'calling or occupation'* was stated as surveyor and architect. The passenger list included such eminent travellers as Lieutenant H. Watkins of the Australian Antarctic Expedition and the Bishop of Perth.

Five months later, on the 12th of January 1912, one day after his 23rd birthday, Charles's friend Waldemar applied for a passport for the purpose of travelling to Perth in Western Australia. At the time, he was living at 16, Strada Nuova, in Msida. Waldemar's signature was authenticated by Dr. Gustavo Busuttil, the Msida district medical officer.

Furthermore, Waldemar's university friend, Anthony, had lodged his passport application two days earlier. At the time, he was residing at 70, Strada Levante, in Valletta. Legal procurator E. Vella authenticated his signature.

In those days, the passport application consisted of an A4 page form containing the personal details of the applicant, namely the full name, birthplace, date of birth, residence, destination, passport number, date of lodgement, and the details of the person authenticating the applicant's signature. There was also space for remarks, which could include such details as the date of arrival at the foreign destination, if such information was available. The fee to lodge the form was around two shillings six pence.

From 1915 onwards, the forms became enriched by the photographs of the applicants. These were often than not larger than today's passport photos and in some cases were group photos of persons travelling together, especially in case of families.

Sydney-based researcher and historian Mark Caruana conducted an extensive study on early Maltese passport applications, which are held at the National Archives in Rabat, Malta. He discovered that the year 1912 saw a marked and

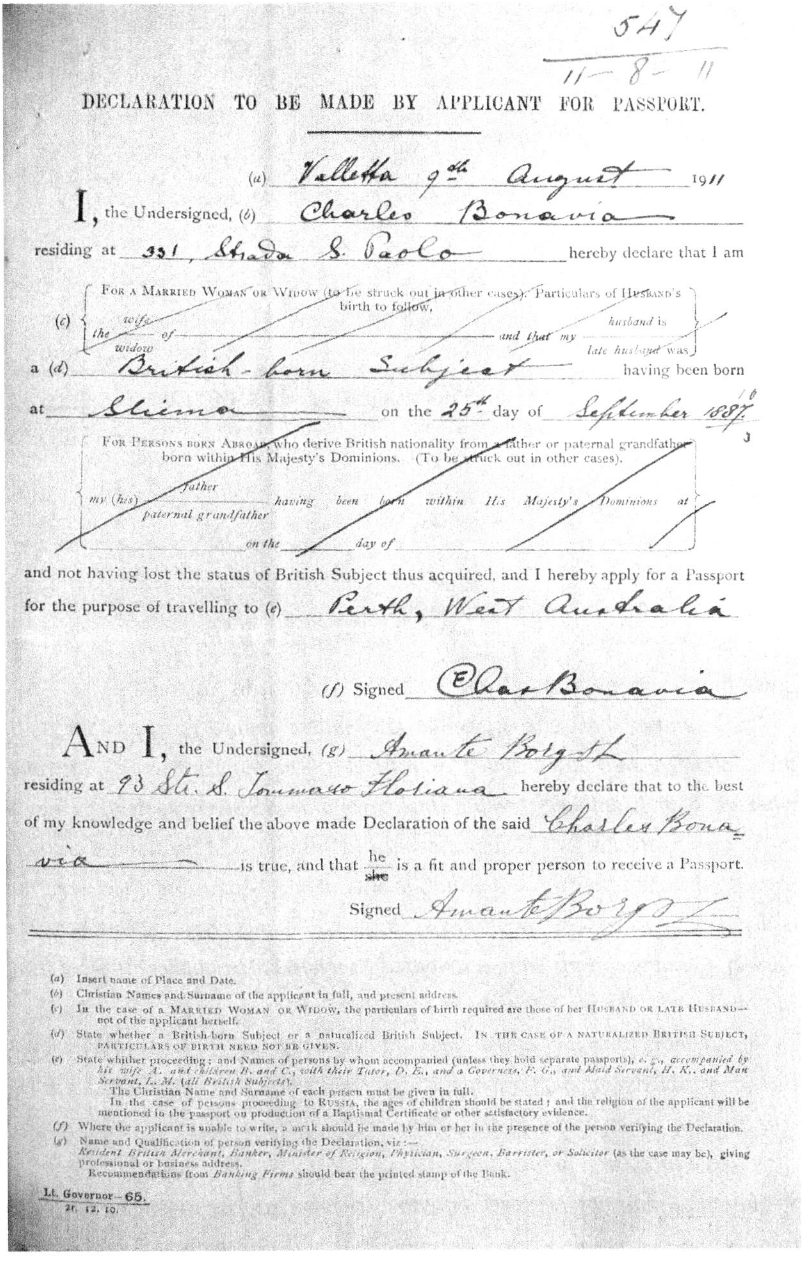

547
11-8-11

DECLARATION TO BE MADE BY APPLICANT FOR PASSPORT.

(a) Valletta 9th August 1911

I, the Undersigned, (b) Charles Bonavia

residing at 331, Strada S. Paolo hereby declare that I am

(c) [For a Married Woman or Widow (to be struck out in other cases). Particulars of Husband's birth to follow.] ~~the wife/widow of ... and that my husband is/late husband was~~

a (d) British-born Subject having been born

at Sliema on the 25th day of September 1887.

[For Persons born Abroad, who derive British nationality from a father or paternal grandfather born within His Majesty's Dominions. (To be struck out in other cases).] ~~my (his) father/paternal grandfather having been born within His Majesty's Dominions at ... on the ... day of ...~~

and not having lost the status of British Subject thus acquired, and I hereby apply for a Passport for the purpose of travelling to (e) Perth, West Australia

(f) Signed Chas Bonavia

And I, the Undersigned, (g) Amante Borg ...

residing at 93 Str. S. Tommaso Floriana hereby declare that to the best of my knowledge and belief the above made Declaration of the said Charles Bonavia is true, and that he ~~she~~ is a fit and proper person to receive a Passport.

Signed Amante Borg ...

(a) Insert name of Place and Date.
(b) Christian Names and Surname of the applicant in full, and present address.
(c) In the case of a Married Woman or Widow, the particulars of birth required are those of her Husband or late Husband—not of the applicant herself.
(d) State whether a British-born Subject or a naturalized British Subject. In the case of a naturalized British Subject, particulars of birth need not be given.
(e) State whither proceeding; and Names of persons by whom accompanied (unless they hold separate passports), *e. g., accompanied by his wife A. and children B. and C., with their Tutor, D. E., and a Governess, F. G., and Maid Servant, H. K., and Man Servant, L. M. (all British Subjects).*
The Christian Name and Surname of each person must be given in full.
In the case of persons proceeding to Russia, the ages of children should be stated; and the religion of the applicant will be mentioned in the passport on production of a Baptismal Certificate or other satisfactory evidence.
(f) Where the applicant is unable to write, a mark should be made by him or her in the presence of the person verifying the Declaration.
(g) Name and Qualification of person verifying the Declaration, viz:—
*Resident British Merchant, Banker, Minister of Religion, Physician, Surgeon, Barrister, or Solicitor* (as the case may be), giving professional or business address.
Recommendations from *Banking Firms* should bear the printed stamp of the Bank.

Lt. Governor—65.
26. 12. 10.

*17. Charles's passport application*

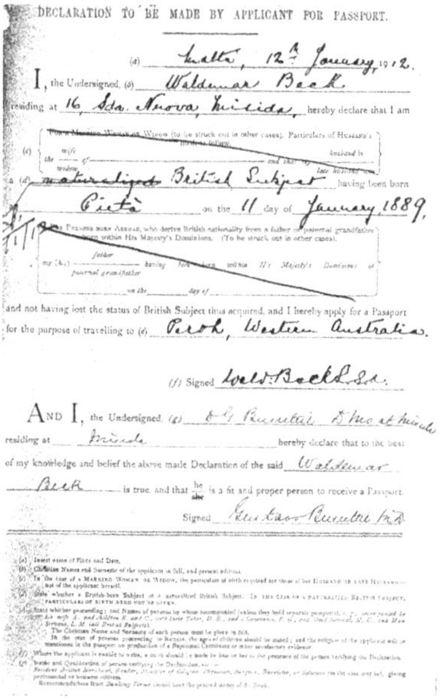

DECLARATION TO BE MADE BY APPLICANT FOR PASSPORT.

(a) Malta, 12th January 1912.

I, the Undersigned, (b) Waldemar Beck

residing at 16, Sda. Nuova, Misida, hereby declare that I am

a (d) ~~naturalized~~ British Subject having been born

Pietà on the 11 day of January 1889,

and not having lost the status of British Subject thus acquired, and I hereby apply for a Passport for the purpose of travelling to (e) Perth, Western Australia.

(f) Signed Wald. Beck

AND I, the Undersigned, (g) Dr. G. Buttigieg, M.D.

residing at Misida hereby declare that to the best of my knowledge and belief the above made Declaration of the said Waldemar Beck is true, and that he is a fit and proper person to receive a Passport.

Signed Gustavo Buttigieg M.D.

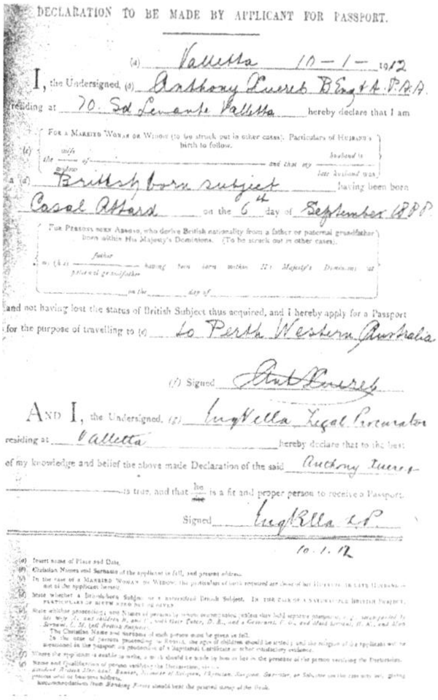

DECLARATION TO BE MADE BY APPLICANT FOR PASSPORT.

(a) Valletta 10-1-1912

I, the Undersigned, (b) Anthony Buttigieg B.Eng & A.R.A.

residing at 70. Sd Levante Valletta hereby declare that I am

a (d) British born subject having been born

Casal Attard on the 6th day of September 1888

and not having lost the status of British Subject thus acquired, and I hereby apply for a Passport for the purpose of travelling to (e) to Perth Western Australia

(f) Signed Ant. Buttigieg

AND I, the Undersigned, (g) Eug. Vella Legal Procurator

residing at Valletta hereby declare that to the best of my knowledge and belief the above made Declaration of the said Anthony Buttigieg is true, and that he is a fit and proper person to receive a Passport.

Signed Eug. Vella L.P.

10.1.12

*17(1). Waldemar and Anthony's passport applications*

quick increase in migration from the Maltese Islands to Australia:

*'By the end of that year, a total of 438 Maltese applied for a passport with their intended destination stated as Australia. The majority wrote 'Australia' whereas 36 applicants specified the State, town or city: 36 were going to Sydney, 4 to Western Australia, 2 to Melbourne and 1 to Tasmania. In the same year, there were 8 other persons whose destination was also Australia but they were not issued a passport in their own right as they were infants or young children. The total of 438 applications were distributed as follows: 416 applicants were from Malta whereas 22 were Gozitans.*

*Slightly more than half of the applicants were illiterate. In lieu of their signature, they placed a cross on the application form which was authenticated as the true mark of the applicant by the person verifying the declaration, who in most cases was a priest or a notary public.*

*Most applicants applied together with a friend or relative for support, usually from the same town or village. Mostly they travelled in pairs or small groups. Blood relatives (including wives and children) would follow some time later.'*

Anthony and Waldemar followed this trend of travelling in pairs. The Malta Customs House passenger list, held at the National Archives at Rabat, for the P&O ship *Simla* shows them sailing from Malta on the 16th of January 1912. Their nationality was stated as English and their profession as architect for Waldemar and engineer for Anthony. In Port Said, they transferred onto the *Marmora*, which, after stopping in Colombo, took them to their final destination at Fremantle on the 13th of February 1912.

***18. Waldemar before his departure***

*19. A photo dedicated to Waldemar by his uncle, Herman Maempel*

Upon arrival, D. McCormack, a medical officer, certified that all passengers were physically and mentally healthy and of good character. This declaration was also signed by the master of the vessel and was made in line with the Immigration Restriction Act:

*'...to the best of my knowledge and belief none of the Passengers in the list (except as therein stated) is an idiot or insane person, or a person suffering from an infectious or contagious disease, of a loathsome or dangerous character, or a person likely by reason of any physical disability to become a charge upon the public or upon any public or charitable institution.'*

*20. Farewell by family and friends*

Charles, Waldemar and Anthony were not the first of their compatriots to try their luck in faraway Australia. After the arrival of the very first Maltese in Australia as convicts in 1810, the first free settler arrived in 1838. In the 1880's, a first attempt was made at an organised system of subsidised group migration to the continent when, in 1883, seventy Maltese travelled to Queensland to take up employment on the sugar plantations. This was the first step in a long and arduous road, culminating in 1948 with the implementation of the Assisted Passage Agreement between Malta and Australia, which brought thousands of Maltese to Australian shores.

The efforts made in the 1880's to put emigration on an organised footing met with little success. In 1882, Francesco Decesare, an elected member of the Council of Government and the proprietor and editor of the Italian language newspaper *'Risorgimento'*, was sent on a fact-finding mission to investigate Australia as a potential country for Maltese migrants. His visit took him to Queensland, New South Wales, Victoria, Tasmania, South Australia, and New Zealand over eight months. He held meetings with governors, premiers, bishops, farmers, the press, and employers who were very keen to employ Maltese, hoping to clinch a deal to set up an assisted passage scheme. He was enthusiastic with what he found and called Australia *'a terrestrial paradise'*. However, his visit was not a great success since the Australian government was reluctant to assist as it feared opposition from the labour movement. Furthermore, the fact that Decesare was acting without authority from the Maltese government did not help his case.

By far, Queensland was the most generous state. It offered to pay half of the passage, with the other half to be funded by the Maltese government. Decesare did not accept this offer as he felt that this would be too much of a burden on the Maltese authorities. At the time, non-British subjects, such as Germans and Scandinavians, were emigrating via London on free passages. Some Maltese farmers and villagers still went on their own accord to work in the sugar cane fields. The Sunshine State also offered land to be bought at a nominal price for the Maltese; however, this offer was also declined by Decesare.

En route from Adelaide, he met the Governor of Western Australia, Napier Broome, and the Mayor of Perth, who invited him to stop at Fremantle.

Governor Broome promised to prepare a Bill to assist Maltese migration to his State. Once again, Decesare refused the invitation.

At every opportunity, Decesare lauded the excellent qualities of Maltese workers: they worked hard and were accustomed to harsh, hot conditions. He mostly encouraged labourers, not professionals, although there were some openings for land surveyors. One of them, Carmelo Fabri from Birgu, had migrated to Adelaide. He started working with the South Australian Survey Department but was laid off. Following this, he ended up selling off his books and instruments as he could not find other work. He became destitute and was admitted to Adelaide's Destitute Asylum. The South Australian government refused to pay for his passage back to Malta as allegedly, he had refused manual work, which he thought beneath him. He told Decesare that he would rather die of hunger than take the menial jobs on offer. Decesare promised him that, upon his return to Malta, he would plead with the authorities and his family so that he be sent money for his trip back home. Eventually, one of Carmelo's relatives travelled to Australia to escort him back to Malta.

Another bad start, which did not augur well for any other prospective Maltese migrants, occurred on the 2nd of October 1883, when sixty-one labourers and nine stowaways sailed for Queensland on board the *Nuddea*, accompanied by a Capuchin priest Father Ambrose Cassar. When they arrived in Townsville, the Maltese were in awe of its prosperity and soon started to enjoy the glitter and glamour of the city so much, they refused to board the ferry that was to take them to the plantations. The harsh, unfriendly land full of poisonous snakes was a far cry from the *'terrestrial paradise'* Decesare had promised! Most of all, they even discovered that they would be paid slave labour wages. Some of them were arrested and sent to prison. By the end of 1884, only sixteen out of the original sixty-one remained on the plantations; others decided to try their luck in Sydney, while the rest returned to Malta.

However, not all proved to be doom and gloom in these early days of Maltese migration. There were also some notable individual success stories. Canadian-born Maltese Frederick William Ponsonby Cammilleri first came to Australia as a seaman in 1879, when he was eighteen years of age. In 1885, he settled in Sydney until he was lured by the gold rush in the Kalgoorlie district, where he became one of the first prospectors. In 1894, Lady Luck smiled upon him when he opened up a shoot of gold on the Brown Hill lease that would make him and the region famous. This first lode-formation discovered in Western Australia was around 1,000 metres in length, 45 metres wide, and 21 metres high. It was the richest discovery in that State, yielding about £7 million worth of gold. Besides his gold digging successes, Frederick was also active in community life and was elected councillor and later a mayor for Busselton. He passed away in Western Australia in 1943, aged eighty-two.

Decesare asked for more time to remain in Australia but his request was refused. Back in Malta, for political reasons, the Maltese authorities censored parts of his report - another nail in Decesare's political coffin. In 1884, he was finally defeated at the polls by the anti-reformists.

In the 1880's, pro-British Maltese newspapers such as '*The Malta Standard*', being staunch supporters of Australia as the ideal destination for Maltese migrants, gave prominence to news from that part of the world. In 1882, another newspaper even went as far as to appoint its own agent for Australasia, based at Lonsdale Street in Melbourne.

Despite these early concerted efforts to encourage the Maltese to emigrate, there was very little emigration from Malta between 1885 and 1911. During this period, only a hundred or so Maltese went to Australia, some of whom were assisted by two special funds.

The Pappaffy Fund was set up by Giovanni di Niccolo Pappaffy, a Greek who had migrated to Malta. Knowing firsthand the hardships encountered by migrants, he set up this fund, which assisted males aged between 18 and 24 with their fares and limited maintenance when they arrived at their destination. Although the original intention of the fund was to assist those of limited financial means, in reality, it was mostly professional people and tradesmen who benefitted. The criteria of literacy and passing a test in two languages made it difficult, if not impossible, for the poor to qualify. Those proficient in the difficult art of calligraphy were also given preference! Between 1892 and 1910, fourteen Maltese were sent to Australia after receiving assistance from this fund. They were educated and middle class and probably gave a good impression as rightful ambassadors for the Maltese! Many more received endowments through the Pappaffy legacy over the years, even as late as the 1950's.

Another fund to assist prospective migrants was bequeathed by Marchese Vincenzo Bugeja. The Church also set up the League of the Apostleship of Prayer and established a Penny Bank for those who wanted to put their money aside to buy their passage. Although these funds proved helpful and beneficial to a number of migrants, a more systematic and organised scheme was required, possibly under the auspices and control of the government. This need became particularly pressing in the early 20th century when the twin problems of overpopulation and unemployment reared their ugly heads.

The 1890's and initial years of the 20th century had brought prosperity and hope for the Maltese, thanks to an economic boom and jobs available for all. The Grand Harbour was buzzing with activity and various projects guaranteed employment and a secure future for the workers. Men abandoned their fields to work around the port. For the construction of the breakwater, between 1903 and 1910, workers had to be brought from Spain and Italy as there were not enough Maltese workers.

With no natural resources, Malta's economy depended solely and precariously on the British use of its natural harbour, which in turn depended on the international political situation. The livelihood of thousands of Maltese workers was intimately linked with how much money the British Exchequer was pumping into the defence of the sea routes, especially the Mediterranean. During the Crimean War of 1853-1856, the harbour teemed with British and French vessels. If one could ever mention the word 'war' in the same breath as 'blessing', then this was a blessed time for Malta! The 1869 opening of the Suez Canal meant another boost to the Maltese economy.

However, in the initial years of the 20th century, the luck of the Maltese started to turn. Malta started to feel the impact of strong competition from other ports bordering the Mediterranean littoral: the French were developing

Algiers and Tunis as efficient shipyards, while Port Said and Alexandria had also been flourishing. Furthermore, the Grand Harbour started to lose business as larger ships were bypassing Malta, whereas previously, every ship passing through the Mediterranean had to stop for coal and other supplies. In 1902, both the naval and military Imperial Garrisons were reduced by six battleships and two battalions. This affected the economy in an adverse way and had long-term repercussions.

In 1905, there were 9,000 men working in the drydocks, but when work dwindled in 1906, once the harbour projects were completed, large numbers were sacked. The square in Vittoriosa used to be full of men waiting around for a foreman to offer them work. The number of beggars on the streets increased. It was common to see a woman holding a baby knocking on doors and asking for a job as a maid.

Malta's population was also on the rise, with the rate of more than 2,000 a year. According to the 1901 census, the population had reached almost 185,000 people. In 1907, the population rose to 208,000, which was not considered sustainable for such a small island.

*21. Waldemar and Anthony on their way to a brighter future in Australia*

Another nail in the coffin for the ailing Maltese economy came in 1912. The British Prime Minister Herbert Asquith, accompanied by the future Prime Minister Churchill, the First Lord of Admiralty, and Lord Kitchener, visited Malta on the 29th of May 1912, to discuss the naval situation in the Mediterranean and the North Sea. A month later, the Mediterranean Fleet started to withdraw from its base in Malta, transferring its ships for the defence of the North Sea and the Atlantic, and leaving only four armoured cruisers in Malta.

It was not just the manual

labourers that were struggling. Between 1909 and 1912, around two hundred students, including Charles, Waldemar and Anthony, received their diplomas and degrees from university. Most of them found it very hard to find suitable employment in the areas they had studied.

A need was felt to combat the rising unemployment and overpopulation and the solution was simple: to encourage emigration. The 'how' proved to be a bit more problematic, especially because this hot issue became the major bone of contention between the two major political factions. The pro-Italian Nationalist Party was vehemently against what it termed *'the exportation of humans'*, arguing that it was in the interest of the British to reduce the civilian population so that Malta would better serve its function as a military and navy base. It claimed that Britain wanted to populate its colonies with Maltese so that it would have more soldiers ready to protect its interests. The Nationalists also resented the foreign workers who had been imported to work in the shipyards and who had stayed behind long after the projects had finished.

A lot of time and energy were lost in debating the language question - the prevalence of the English language versus the Italian language and culture. The pro-Italian newspaper *'Malta'* even propagated and supported the idea of annexation by Italy! The real issues that affected the people, such as education, health, and work, were never discussed. The masses could only understand and speak Maltese and they did not care much for English or Italian; their main worry was to ensure that they had food on the table for their families. Yet, they were left to fend for themselves, trying to find from where they could earn their next shilling. It was a tough time for workers, who had two choices: either emigrate or readjust to lower wages.

All the way from Perth, Charles was following closely these events unfolding in Malta. Although far away from his country of birth, he always talked about Malta and took an interest in what was going on. His strong sense of patriotism never left him, and is evident in this extract from a letter to his father:

*'I think that all my friends here have heard enough about Malta and by the way I describe Malta to them, they all must think that it is the Garden of Eden, not that I exaggerate at all.'*

In its edition of the 15th of September 1913, *'The Daily News'* in Perth

*22. Greetings from Perth*

published a letter from Charles who was writing in reply to the column *'At a glance'*. The newspaper in the said article had referred to the 108th anniversary from when Malta had *'capitulated'* to the British forces. Charles disagreed with the use of this term based on the fact that Britain and Malta were never at war with each other. He stressed that the Maltese had voluntarily assented to be part of the British Empire. We recall that his ancestor Matteo Bonavia had played a role in this important historical event.

From his observations, Charles had started to feel that Malta's future was under threat. In a letter to his sister Laura dated March 1912, he expressed his preoccupation when the British fleet withdrew from Malta as its base, saying that he was shocked by the news as this would mean hardship and mass unemployment: *'Unless the people there leave by hundreds a week, they will all starve.'*

In the meantime, the debate on emigration raged on. Historian Dr. Barry York identifies three key players: firstly, the British Colonial Office wanted to solve the problem of Malta's overpopulation. Then there were the Maltese 'emigrationists', whose pro-British fervour led them to encourage an organised system of emigration, as opposed to the previous haphazard attempts. The third player was the Australian government, which, while desiring to send migrants to vacant and remote parts of the country, such as the Northern Territory, still imposed restrictions on Maltese migrants. On the other side of the fence, the anti-reformists (pro-Italian) spread propaganda against Australia, and promoted migration to North Africa and South America. After all, there were already well-established colonies of Maltese in Egypt, Libya, Tunisia and Algeria, and there had also been a free emigration scheme for Brazil. The anti-reformists opposed the increase of British influence in Malta, especially the flourishing of the English language, as they argued that this would lead to Protestantism. Italian was considered the language of the elite and as such was used in the realms of law, trade, and commerce. Only in 1934 did the Maltese language become the official language of the law courts.

A number of Maltese businessmen also opposed emigration beyond the Mediterranean as they had a vested interest in the land they owned, which they rented out to farmers. They feared that, if farmers started to emigrate, the cost of labour would go up, while the rent and prices would drop. Other affluent Maltese took a different perspective: for them, emigration was seen as the perfect solution to get rid of fellow Maltese who could cause trouble and disrupt their comfortable life and status quo. In this, they were supported by the British colonial government who did not care much for the plight of the suffering populace. Rather, what really concerned the government was that some people were listening to revolutionary ideas disseminated against it, which was undermining the colonial government's stronghold. Immigration was the perfect solution to get rid of these dissenting and troublesome voices. One such person, considered a thorn in the side of the authorities, was Manwel Dimech, a radical who preached about the importance of the Maltese language, the emancipation of women, and the evils of colonialism. When the war broke out, he ended up in exile in Egypt, where he would die in 1921.

The reformists were in favour of emigration to English-speaking countries, especially to Australia. They argued that the Maltese had a lot in common with Australians; namely, they were white and both countries owed allegiance to the British flag. They encouraged the learning of English as this would help prospective migrants. Besides preparing the Maltese for emigration to English-speaking countries, the policy of diffusing the English language in Malta had another purpose: to counter pro-Italian sentiment. However, to complicate matters further, the pro-British immigration group was divided as to what emigration scheme and method to adopt.

It was against this backdrop of a precarious economic situation and heated discussions and dissention that the Malta Emigration Committee was formed in 1907, with the aim of working on an organised system of migration. It was a consultative body, independent from the Council of Government, set up to give advice to prospective migrants, and make contact with immigrant countries regarding opportunities for work. In 1910, it suggested all kinds of assistance to migrants, such as loans, assisted passages, and the recommendation that a doctor and a priest accompany the group, but the government did not take heed. Furthermore, the committee only had an advisory capacity and thus, in reality, it did not carry any weight.

An obvious choice for the position of Secretary of the Committee was Charles Mattei. Born in Malta in 1864, he always cut an impressive figure with his sturdy appearance, thick dark moustache and pipe in hand. His sense of adventure took him to Western Australia in 1892, where he planned to invest in the mining industry. In 1896, he was attracted by the gold rush and tried his luck at the Kalgoorlie goldfields. Mattei had completed his medical studies in Glasgow and Dublin. With this background, he organised the St. John Hospital and became a pioneer medical practitioner on the fields. In 1900, he was appointed Medical Officer and Resident Magistrate at Wyndham and Government Electoral Registrar, as well as Protector of Aborigines. His connection with Tasmania included a share in Australia's first electric trams in Hobart and the Mount Zeehan silver mines. He was perhaps one of the very first Maltese who travelled to China. In 1901, Mattei served as a Medical Officer of the Australian Army Medical Corps during the Boer War in South Africa, serving with the Tasmanian Bushmen.

Mattei returned to Malta in 1906 where, due to his experience in foreign lands, he presented himself as an official emigration agent representing the Maltese government. As a practical way to help Maltese migrants whose English was not up to scratch, he used to provide a written notice showing the person's name, nationality, destination, and a request for help. He also suggested that they wear a Maltese cross so that whoever was waiting for them upon arrival would recognise them instantly. In a letter dated 17th of March 1911 to a local newspaper, Mattei, in describing this procedure, likened these migrants to postal items:

*'Practically I attached a label to this man and sent him by mail carriage paid and he was safely delivered.'*

*'There is now a 'parcel' of six women and four children waiting to be 'mailed'.'*

In 1912, the Emigration Committee sent Mattei on a fact-finding mission to British Columbia, the United States and Australia. A number of Maltese were already working with the North Pacific railway along the Canadian border, and he recommended California and British Columbia as ideal places for Maltese to settle in.

During the First World War, Mattei volunteered and was assigned to No.1 Australian Stationary Hospital. He left Australia in 1914 with the first medical units and served in Mudros, Gallipoli and France. In 1916, he rose to the rank of Major and two years later, to Lieutenant Colonel. After his discharge in October 1919, he went to London hoping for an appointment in the Colonial Service. Disappointed that this did not materialise, he returned to Australia where he became the Acting Superintendent at Queen Alexandra Hospital and the Tropical Diseases Clinic at Rockhampton.

The 1920's saw Mattei returning to Malta but he was back in Australia in 1933. He toured the sugar districts of North Queensland, promoting himself as the ideal candidate for the position of Malta's Commissioner in Australia, to replace the non-Maltese Commissioner Frank Corder. In 1934, he also formed an Australia-wide Maltese-Australian Association, which, unfortunately, did not gain much ground, largely through the hostility of the prosperous Maltese in North Queensland.

He returned to Malta where he kept an interest in Maltese emigration and, at age seventy, he stopped travelling and relied more on the mighty pen to recommend new settlements for the Maltese. A case in point was when he proposed a Maltese settlement for cotton-growing on the Dawson River Theodore Irrigation Area to the Queensland government. The proposal was not accepted due to the Australian government restrictions on Maltese immigrants in the 1930's. Mattei passed away on the 20 August 1939, just a few days before the start of the Second World War.

Another member of the Emigration Committee was Professor Lorenzo Manche`, who replaced Mattei when the latter was on his many travels. Manche` had three main suggestions. The first was that issues relating to emigration should fall under the remit of private enterprise, the church, or the Colonial administration. He held the view though that such matters would be best left in the hands of the government. Manche` also recommended that a representative would go where Maltese intend to migrate to lay the groundwork and then help them settle. Another of his proposals was that a Maltese priest and doctor would accompany the migrants. He also championed the idea of entering arrangements with the main shipping companies of the time to provide free passages to *'bona fide'* persons. Manche` opposed migration to Canada and Australia as the former was too cold and the latter was too English!!

Professor Manche`, who had an ophthalmic clinic practice in Ħamrun where he examined prospective migrants, was aware that a number of Maltese were repatriated to Malta from the country of migration because of health issues. He stressed the importance of sending people with a clean bill of health. At the time, lack of hygiene brought about a slew of diseases among the Maltese population, such as tuberculosis and Malta fever. Trachoma, an eye disease that could lead to blindness, was also widespread, due to the unhygienic and dusty atmosphere. In November 1911, cholera struck Malta, brought in by Maltese refugees who had left Tripoli because of the fighting between Turkey and Italy in Libya. By December, it had killed 85 people.

In the same year, a contingent of Maltese migrants was not allowed to land on Ellis Island, New York. After much sacrifice to raise the funds for the passage and the long journey, they were obliged to go back to Malta as their eyes were infected with trachoma. From 1912, a health certificate became a prerequisite to disembark in Australia.

The Royal Commission on financial, economic, and judicial procedure headed by Sir Francis Mowatt met between August 1911 and January

1912. One of the areas tackled was the issue of migration. The Commission recommended emigration on a large scale as the only remedy for the island's overpopulation. It stressed that this should be officially encouraged, especially within the British Empire. In particular, the northern territories of Australia were considered ideal. Loans from the government and medical tests to prospective migrants were proposed, as well as the recommendation for the establishment of an Emigration Information Office, and the creation of a Greater Malta concept overseas, whereby land would be bought where pockets of Maltese could be established. Emigrants had to be carefully selected and adequately prepared for the different conditions of the receiving countries. The Maltese government had to take a proactive approach to emigration and place it on an official footing. In 1914, the Special Dominions Commission, whose mission was to plan the development of the Empire, also came up with the same recommendation.

These recommendations looked good on paper, but when it came to implementation, it was another matter. The Maltese did not have any political clout, and therefore whether these suggestions would bear fruit depended solely on the willingness of the Colonial government. But, the government was not ready to fork out any money to finance people who needed to emigrate, and it did next to nothing to implement the policy it advocated. Malta was considered solely for its strategic value, for as long as it was serving the interests of the Empire well. The well-intended efforts of assistance by benefactors Giovanni di Nicolo Pappaffy and Marchese Vincenzo Bugeja were small when compared to the need, as most people did not have sufficient funds to cover the passage. Various attempts were made at raising funds, including suggestions to organise balls and concerts as fund-raising activities.

The newspaper *'The Daily Malta Chronicle'* had also attempted to organise an Emigrant Bank, which was not a great success. Despite this failure, the newspaper proved to be very influential in the debate on emigration, and a staunch promoter of an organised and subsidised system, especially for educated middle-class young men. The newspaper supported systematic emigration, which would avoid the wrong impressions of who the Maltese really were. This guaranteed against the sending of people to places where they were not wanted and created a 'correct fit': organised emigration '*would direct our surplus population to the land that would be best for us and for which we would be best.*'

'*The Daily Malta Chronicle*' dedicated various editorials to the issue and published several letters by Maltese persons living in Australia. Most of them came from the '*Southeastern part of the Commonwealth*', that is, '*the first regions colonised and the most thickly populated.*' These areas were described as advanced and developed, but the newspaper stated that it was harder for migrants to settle in them unless the Australians themselves '*have directly or indirectly invited them to come to them.*' The newspaper seemed to favour Western and North Eastern Australia as '*a suitable place of settlement for our surplus population than Melbourne or Sydney.*' Western Australia was said to have great promise and potential. One of its advantages was that '*its climate resembles our own.*'

By this time, Charles had settled well in the Perth community, after the normal teething problems of any newly-arrived migrant. He was keen to join the debate on migration, speaking from firsthand experience, and holding firm views on the lack of planning by the Maltese authorities in sending illiterate people to Australia.

One of his letters to his father was published in '*The Daily Malta Chronicle*' of the 30th of May 1913. He wrote: '*I still am a keen follower of all correspondence about emigration. I am glad to see that at last the Maltese man in the street is looking deeply into this subject. I am still more glad to see that the majority of the emigrants are going to other portions of the British Empire.*'

He considered the issue of migration to be '*perhaps the greatest of our present economic questions.*'

'*It would be a loss to Malta and the Empire if such good men were allowed to go to foreign countries. As it is, at least another portion of the Empire is gaining at Malta's expense. I referred to the emigrants as 'good men' because I can judge by the people we have here and by what I read of others who have emigrated to other places.*'

Charles was for a selective system where only those who were fluent in English could migrate to Australia. Regarding the bad impression some

Maltese were giving, he put the blame squarely on the authorities:

*'If fault was found with them because of the way they dress and because they cannot speak English, it is simply the fault of the local authorities, who, if unable to raise funds to help emigration financially could at least organise a sort of 'Information Bureau' and also a sort of committee to select men out of the willing emigrants and give them a letter of introduction to the authorities of the places the men are going to.'*

The picture of a group of uneducated and unskilled Maltese migrants is painted vividly in the satirical novel *'Ulied in-Nanna Venut fl-Amerka'* (*'The children of Grandmother Venut in America'*), published in the 1930's. The author, Ġwann Mamo, tells the story of a motley crew of Maltese, including himself, who embark on the long journey to the United States of America. Most of them are illiterate, with no knowledge of even the most basic English words. Their adventures, or rather misadventures, on the way and once they arrive in a big American metropolis, are recounted in half humorous, half tragic tones. The book was intended as a dig at the authorities in Malta that did little to improve the plight of the low strata of society and thus, sent the Maltese away without sufficient preparation and support.

One must remember that the vast majority of these early migrants had never before stepped out of the borders of their little village, let alone beyond Malta, and now they were making their way on their own into the big, bad world.

Once again, Charles pointed the finger at the government for Malta's problems:

*'Although Malta is over-populated, if things were carried on there, as they are in other parts of the world, the island, small as it is, would be big enough to hold the lot of them.'*

The newspaper supported Charles's views with an editorial:

*'We have before us, as we write, an extract of a letter sent from Western Australia by Mr Charles E. Bonavia B.Eng and A LS and A. It was addressed to his father Mr E.T Bonavia of the Malta Civil Service. The writer of the letter again evinces the greatest practical interest in his Maltese fellow countrymen, and the island of which he is as proud as ever, and he makes proposals which it might be well to adopt and put in practice.'*

*'We like to turn away from discord and railing and think a little of Mr Bonavia in Western Australia and of the suggestions he makes.'*

The newspaper had words of praise and admiration for Charles:

*'Mr Bonavia went to Australia to practise his profession. He had his difficulties as the degrees of our University are not recognised in the Commonwealth. But he was not daunted by these difficulties. He prepared himself to triumph over them with that spirit which carries a man always through his troubles. His position is now assured: he looks with confidence to the future.'*

Continuing to comment on Charles's letter, the editorial opined:

*'Our folk have gone away, driven by their misery. Fit and unfit we have with indifference let them take their departure. We have suffered them to leave Malta often in an unpresentable state. We have in many cases been content to be represented by persons from amongst us who hardly even represented themselves in their ordinary conditions. We have done Malta much harm by all this.*

*We have closed doors which we ought to have striven to keep wide open. Mr Bonavia does not speak of systematic emigration, but he has organised emigration in mind. If we are able to start the proposed 'Emigration Bank', the evil committed may be in part, or wholly repaired.*

*The parts of the Empire best suited to receive the excess of our population ought to have been prepared to receive them. We have not yet determined to what colony emigrants had best go, all things being taken into consideration. We have been neglectful of the interests of Malta and of those of our countrymen who leave Malta.*

*The letter of Mr Bonavia recalls us to a sense of our duty. That duty is somehow summed up by implication in the declaration that the Maltese are good men for the Empire and that the Empire is a good place for them.'*

Charles warned that the Maltese '*must be aware and not spoil their chances in Australia.*' He reiterated the importance of giving a good first impression to employers and fellow workmen. He also mentioned that carpenters, bricklayers, general labourers, and land clearers were in great demand, earning between ten to twelve shillings a day. The State Records Office in Perth holds Waldemar's service card, showing his temporary employment as a draughtsman with the Western Australian Government Railways. His salary rate *per diem* was ten shillings. He managed to secure this job just two months after he arrived, and although it only lasted until June 1912, it would have given him the necessary confidence and experience, which would come in handy when applying for more long-term work. At the time, in Malta, the average wage was fourteen shillings a week, but most workers could only dream about earning this amount of money. It was hard to raise a family, which in those days consisted of numerous children, on that miserly sum.

Reports sent home by Maltese migrants were favourable and full of encouragement to their friends and families to join them. When a letter arrived from Australia, there was great excitement and commotion. These letters, which invariably were taken to the local parish priest to be read, due to the high illiteracy rate, included gifts of £30-£50, considered as huge amounts of money at the time. News of these much needed windfalls spread like wildfire and served as the best incentive to encourage emigration and keep the debate going. The talk in the church square, in the local bars and band clubs was that whoever migrates '*isir nies*', meaning that one becomes prosperous and successful.

Echoing the newspaper's emphasis on careful selection and a sound knowledge of English as a prerequisite expounded in the 8th of May 1913 edition, Charles stated that three criteria had to be satisfied: the person intending to migrate to Australia had to speak English, he must also be ready to work hard, '*not be disposed to shirk work*', and be of good character.

Charles pleaded directly with the authorities:

*'But don't send men who are not reliable. I don't want to recommend men who will turn out to be loafers.'*

He stressed that persons planning to migrate needed to have money set aside to live on once they first arrive. Charles even went as far as to promise that, if anyone goes to Australia without any money, he was willing to fork out his own money to cover their board and lodging, so long as they would pay him back once they started working. He was more than ready to help those who wanted to help themselves but he could not stand people who were lazy and who did anything to avoid work and responsibility.

In another letter to his father, Charles again expressed his readiness to help newly arrived migrants: '*We are ready to help them. Only impress the conditions on them.*'

Paul Calleja in his book '*Maltese of the Western Third*' mentions how Charles became a leader of Perth's small Maltese community, using his contacts to help other Maltese find work. Sometimes, if it was not possible to find them jobs, he employed them temporarily himself, or referred them to the Bishop of Perth, Patrick Clune, who was a great friend to this early community of Maltese pioneers. He assisted Charles in placing new arrivals in employment, and he also managed to obtain £100 from the State government to pay for their lodgings. They all found good jobs, some earning fourteen shillings a day. During the war, his office corresponded with soldiers at the Front, including Waldemar who, in one of his letters to his brother, mentions that he received a card from the Bishop's chauffeur: '*The card went to WA and the Gugeris sent it over here. I mean the Bishop of WA, Dr Clune.*' (Letter to George dated 18th of March 1917)

Bishop Clune would later play an important role as an intermediary between British Prime Minister Lloyd George and the Irish leaders, during the Irish War of Independence.

Besides the Bishop, Charles also called upon his friend, Dr. Bartolo, and Father Patrick Joseph O'Grady to help the new migrants. Charles frequently corresponded with Father O'Grady, the pioneer of the Salesian Order, who had arrived in Malta in 1903 to open a home in Sliema for poor and abandoned boys, some of whom he wanted to send to Australia. He celebrated the silver jubilee of his priesthood in Malta in 1915 and later, in 1928, his superiors appointed him as Director of the Melbourne Salesian College at Rupertswood.

When Charles was joined by Waldemar and Anthony, the three friends did all they could to encourage Maltese settlement in Western Australia. They acted as mentors to newly arrived migrants and between them, they employed seven Maltese. Two of the three who worked with Waldemar turned out to be the loafers Charles could not stand and thus, they were left to their own devices.

At the time that Charles, Waldemar and Anthony arrived in Perth, in September 1911 and February 1912 respectively, Maltese-born Sir Gerald Strickland was Governor of Western Australia. This fact augured well for prospective migrants from Malta to this State as Strickland was an avid supporter of Maltese migration 'down under'. Soon after his arrival in Perth, Charles had called upon Lord Strickland and Bishop Clune. In one of the numerous detailed and informative letters to his friends and family, he admitted that he felt a bit ill-at-ease talking to them, but '*I can get on allright, I think.*'

Strickland was born in Valletta in 1861, of a Maltese mother, Louisa Bonici Bologna, and a British father, Captain Walter Strickland. After pursuing his studies in Malta, England, and Italy, he played a pivotal role on the political scene in strengthening the bond between Malta and the British Crown. In 1883, he toured Australia and his findings supported those of Decesare, recommending Queensland as the ideal destination for Maltese labourers. He was all in favour of Maltese migrating to Australia, a British domain, and in strengthening the influence of the English culture and language in Malta, as opposed to the Italian language. Strickland urged the Maltese to go to Australia by personally writing letters to former colleagues and friends. Few knew that he advanced his own money without interest and anonymously to a number of Maltese migrants.

As Governor of Western Australia between 1909 and 1913, he encouraged young graduates to migrate there, perhaps striking a chord with their appetite for adventure, and the excitement of new faces and places. Charles was one of the first who answered the call. Despite the fact that degrees from the Royal Malta University were not recognised, he was very lucky as, shortly after his arrival, with the support of Governor Strickland and Bishop Clune, he became assistant surveyor of Crown lands, with a salary of £20 a month, a princely sum for those days. The courtesy call he had made to these two influential people, shortly after his arrival in Perth, had paid dividends!

Strickland's political career had started in 1886, when he was elected to the Council of Government and in 1889, he became the Chief Secretary of Malta. His career then took him to a number of foreign posts, namely as Governor of the Leeward Islands in French Polynesia in 1902, and Governor of Tasmania from 1904 to 1909.

While in Perth, his wife Lady Edeline became very ill with an undiagnosed disease of the spine. By autumn 1912, his eldest daughter Mary had assumed the role of hostess at Government House, instead of her ailing mother.

In 1913, Lord Strickland took up the appointment of Governor of NSW, until 1917. Upon his return to Malta, he became an even more prominent player on the local political scene. Following the granting of self-government in 1921, he formed the Anglo-Maltese Party, which shortly afterwards merged with the Maltese Constitutional Party, and became known as the Constitutional Party. Strickland was the leader of the Opposition from 1921 to 1927. In 1927, his party held the majority of seats with the Labour Party in the Legislative Assembly, and he became Head of Ministry.

Between July 1932 and November 1933, Lord Strickland was again the leader of the Opposition and, after the new Constitution in 1939, became the leader of the elected majority in the Council of Government. He died on the 22nd August 1940 at his residence, and is buried in the family chapel at the Mdina Cathedral.

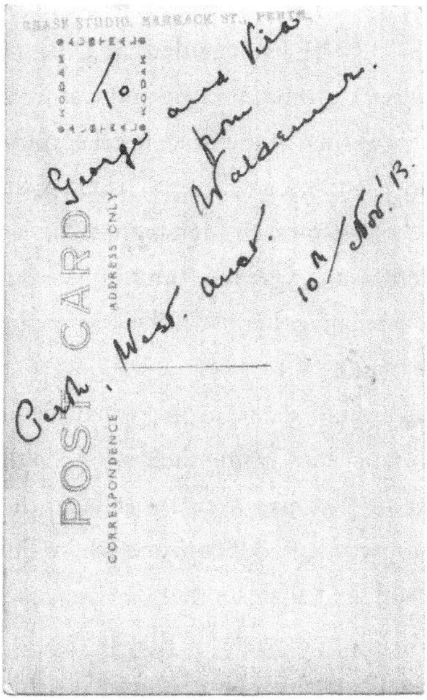


*23. Postcard to George and Vira*

Major General Sir Harry Barron, who had set up the Malta Emigration Committee in 1907, succeeded Strickland as Governor in both Tasmania and Western Australia. In Malta, he administered the government during the Governor's absence and was the Commander of the Royal Artillery in Malta.

Barron was another champion of Maltese emigration to Australia and, together with Strickland, promoted the Maltese as the ideal immigrants, highlighting their good qualities, such as devotion to their families, hard-work, adaptability, and loyalty to the Empire and to Christian and European values. They also had big families which augured well for a population increase in Australia! In a letter to Vira, Waldemar alluded to Australia's issue with its low population and mentioned a scheme whereby the government paid £5 to every new mother, even if the child was born out of wedlock:

'*...Kieku hawn il-gvern kien itih £5. Kull persuna li titwieled, anke illeġittima, ommha tirċievi £5. M'għandhomx nies għalhekk għal kwalunkwe kost sabiex ikollha nies aktar.*' (Letter dated 14th of May 1916)

Another event at this time, that was seen to have helped in the push to facilitate Maltese migration to Australia, was the 24th International Eucharistic Congress held in Malta between the 23rd and 27th of April 1913. Although only 3,000 people attended from overseas, a few of these were some of the most eminent and influential people at the time. The VIP list included the Duke of Norfolk, and five cardinals - among the latter, Cardinal Bourne and Cardinal Ferrata. The reason for the low attendance was that, at the time, hotel accommodation in Malta was scarce. Furthermore, a number of people would also have not cherished the idea of crossing the seas to reach Malta, especially as the situation in Europe, particularly in the Balkans, was very tumultuous and unstable. The bout of cholera, that a year and a half earlier had killed 85 people, also made people wary of travelling to the island.

The Congress ended with a procession of around 10,000 people who walked through the main streets of Valletta to Floriana, showcasing their strong faith in the Catholic Church. Little did they know that, just over a year later, the world would find itself in the middle of unprecedented bloodshed where the unthinkable would happen: Christians would fight against other Christians.

From Australia, Waldemar had written to his mother, saying that an Australian bishop would visit her. According to the Beck family lore, recounted to me by Waldemar's great-nephew, Carolina invested in a new set of shining cutlery for this very special occasion. The bishop in question was the Bishop of Perth, Patrick Clune, who led the Australian delegation comprising of 400 pilgrims. Another representative was the Member of Parliament G.P. Connolly, the former Colonial Secretary in the legislative assembly of Western Australia. Both the Bishop and Connolly had a high interest in migration and believed in its benefits for Australia.

Strickland saw the presence of these two key people in Malta as very timely and as a golden opportunity to promote emigration to Australia. Bishop Clune and Connolly delivered a public lecture in Valletta, encouraging skilled workers and farmhands to Western Australia, while saying that clerical workers were not in high demand. They also addressed a special meeting of the Emigration Committee, praising Strickland and offering their active support to future Maltese immigration to Australia. Bishop Clune pledged the support of the Catholic Church, which would provide land to set up a small colony of Maltese settlers.

If Strickland was happy with the fruitful presence of the Bishop and the Member of Parliament from Australia, he was dismayed with the attitude of Major John Eugene Clauson, who had become Lieutenant Governor and Chief Secretary of Malta in 1911. Strickland had encouraged Clauson to take advantage of Bishop Clune's visit to Malta. However, the major proved *'over cautious'* on the question of helping Maltese to migrate to Australia.

Clune and Connolly spoke highly of other Maltese they knew, mentioning *'gentlemen settlers such as Charles Bonavia, Anthony Xuereb and Waldemar Beck.'* They reported that there were about a 100 Maltese in Western Australia, and they were all doing well. The newspaper *'The Daily Malta Chronicle'* later reported on this:

*'Mr Bonavia is well known to His Lordship Bishop Clune who took part in the Eucharistic Congress held here about a month ago. The latter spoke of our countryman as a gentleman of ability and energy. Bonavia was perhaps the very first Maltese who set foot in Western Australia and he could help now many other to settle there, and share there in degree in his own success.'*

Connolly admitted the prejudice of certain Australians towards Southern Europeans. To counteract this, he encouraged the Maltese to migrate in small numbers so that their presence would not be so conspicuous.

Bishop Clune also met Maltese priests to encourage them to accompany Maltese migrants. One of them, Father Raphael Pace, took up the offer straight away as he went back with the Bishop and Connolly to Perth, where he became Bishop Clune's secretary and looked after the spiritual needs of the small Maltese community there. In 1919, he was appointed the first parish priest of South Perth. The highlight of his career was the christening into the Catholic Church of an Anglican, Ernest John Goody, in 1916, who later became Archbishop of Western Australia. Father Pace was much loved by both Maltese and Australians alike. Incidentally, Father Pace's brother, Francesco, was a member of the Emigration Committee. After a brief return to Malta in 1926, Father Pace remained in Western Australia, until his death in Subiaco in 1953.

The 1911 Australian census reported 248 Malta-born residents: 112 in NSW, 43 in Victoria, 43 in Queensland, 32 in Western Australia, 14 in South Australia, and 4 in Tasmania. The year 1912, during which Waldemar and Anthony took the big plunge, saw a marked increase in the number of Maltese arrivals in Australia. In fact, in that year, 438 passports were issued in Malta for prospective migrants to Australia, whereas only 116 had been issued in the previous year. Between 1911 and 1919, nearly 2,000 Maltese adopted Australia as their new home - almost 1% of the Maltese population at the time. In 1915, more Gozitans than Maltese migrated as the war did not benefit Gozo as much as it did Malta.

Maltese were found working with the NSW Railways on the construction of the North Coast line. Others were in Mackay working in the sugar fields. A hundred or so Maltese worked in the mines and farms of Western Australia. Another two hundred worked in the copper mine owned by the Mount Lyell Mining and Railway Company near Queenstown in Tasmania. In fact, prior to the war, the Maltese were the largest non-Anglo-Saxon group of the mining company's workforce.

The Maltese were accepted in Queensland and in the Northern Territory, where the Australians did not really want to go to work in some of these regions, which were considered as inhospitable.

In 1913, four hundred Maltese signed a contract to work on a project to build railway lines between Pine Creek and the Katherine River, in the Northern Territory. They were to be given a block of land if they decided to stay. Anthony, who was in charge of the Maltese railway workers, was told by the chief engineer of the project that he had never seen such hardworking people. After spending some time in the wheat lands near Wagin, a town some 240 km south east of Perth, Anthony had moved to Darwin where he secured a job at the Lands and Survey Department, and became a member of the Northern Territory Public Service Association. His adventurous spirit also took him to Sydney, where he hoped to find a ship bound for the United States.

Deric Thompson used to work with Anthony in Darwin in the 1930's and 40's. In records held at the Northern Territory archives, Thompson talks extensively about his colleague:

*'One of our great people – I shouldn't say great people; but one of the most popular people in Darwin was a man named Anthony Frederick Xuereb. He was Maltese, and his father was, I understand, Chief Medical Officer of Malta at one stage. And as such Tony had a great knowledge of English naval tradition. He was a great traditionalist and knew – great reader. He was a draftsman. Actually he went to university in Malta, I think as a Civil engineer, but they wouldn't recognise his qualifications here.*

*I can't remember how he came to Darwin; he must have got a job in Darwin, just before the First World War. He got a job as a draftsman. A lot of Maltese came to Darwin about that time, as did a number of Greeks. Perhaps because of Vesteys, I don't know.'* (The Vesteys had a food production business with a meat processing plant in Darwin.)

The same archives hold the transcript of an interview with Arthur Miller who became a surveyor in 1917. He worked with Anthony and also recalled him with admiration:

*'...A mine of information was a draftsman who came there to Darwin in 1912, with a recommendation from the Governor of NSW, he was a Maltese.*

*This gentleman's name - he was a qualified draftsman- was Tony Xuereb...... He was very brilliant...especially on geometric work.'*

In Darwin, Anthony rekindled his friendship with Eligio Castaldi, a law graduate from the University of Malta. Eligio, the son of a University professor, had migrated to the Northern Territory in 1912.

Eligio's son, Professor Peter Castaldi, who lives in Sydney, recalls Anthony (or Tony, as he was known) as a very charming man who was a close friend of his father:

*'Dear Gioconda,*

*How nice to hear from you, made me quite excited. Tony Xuereb was a close friend of my father. I remember him well as a young person as he visited us in Sydney on his leave from Darwin. You are right about them being in Darwin together, Eligio was clerk in the court there and did circuit duties I think in the period 1914-15 after which he went to Suva before returning to Sydney in 1917.'* (Email dated 6th of November 2009)

*24. Eligio Castaldi*

Before Darwin, Eligio had first arrived in Sydney, where he was earning £10 a month. After working in Darwin for nearly two years until 1915, he went to Fiji to take up an appointment with the Imperial Civil Service, until 1917. He served in the war as a corporal with the 54th Battalion, and upon his return to Australia in 1920, practised the legal profession in Sydney. Eligio's interest in migration continued over the years as in 1954 we find him delivering a speech to the Catholic Historical Society on the history of Maltese migration.

In 1912, Dr. Richard Arthur, a prominent Australian politician in the NSW Parliament and President of the Immigration League of Australasia, wrote to the Maltese authorities, recommending Australia as the ideal destination for Maltese migrants. Arthur had visited Malta and seen firsthand how hard working and loyal Maltese workers were. In a speech in Parliament, he said that the Maltese can work under a strong sun. He was all in favour of migrants from Malta, in particular to the Northern Territory, in the coffee planting and fishing industries. Southern Europeans were seen to be perfect for this part of Australia as they would easily adapt to the warm climate. Their presence would also counteract the settlement of Asians, which was labelled as the *'yellow peril'*.

The pro-British political faction in Malta had pushed for Australia as a prosperous country and a land of opportunity. In line with Australia's efforts, especially in the latter part of 1912, to populate its vast lands, the push was for Maltese to migrate to regions that were not already established or settled into by Anglo-Saxons. These areas were largely inhospitable and undeveloped. However, most Maltese who wanted to migrate were unskilled and more lured to the big cities, rather than the outback regions of Australia.

The Maltese could rely on the support of men like Dr. Arthur, Bishop Clune, and Connolly who held them in high esteem. However, not everyone welcomed the Maltese with open arms. All was not rosy for these early pioneers, especially in the big cities. They were not allowed to join a union, and if they were employed, the other workers would stop their work in protest. From Western Australia, Governor Strickland begged his friend, Maltese Chief Justice Joseph Carbone, not to let any Maltese go to Australia until they were admitted on equal terms in the unions by their fellow workers.

Australia was advocating a *'whites only'* immigration policy to ensure a homogeneous population and to continue to strengthen ties with the mother country. This could not have pleased Britain more as this nation was eager to extend its influence and populate its empire for defence and economic reasons. Yet, the labour movement and a number of Anglo-Saxon supremacists did not give much weight to the fact that the Maltese were British subjects, like themselves. The unions opposed immigration in general for labour scarcity served workers as it helped to raise pay, and enabled improvements in working conditions.

The Australian Workers Union (A.W.U) went one step further, targeting the Maltese specifically by passing a resolution condemning '*the immigration of cheap Maltese labour*' into Australia. In one of its publications, the A.W.U went as far as to describe the Maltese as the '*Black Menace*'. The Australian government bowed to their pressure and, in 1912, excluded the Maltese from benefiting under assisted immigration schemes. At the time, a trip from Malta to Fremantle via a European port cost around £12 to £16. This expense was certainly a disincentive for those who wanted to migrate, the majority of whom were illiterate, unemployed men. They had to either borrow money, or sell their meagre possessions, to obtain the necessary funds for the passage. Assisted passages were awarded to Finns, Swiss, and Norwegians, who did not even form part of the British Empire, but not to the Maltese. The Maltese, although white and British subjects, were considered to be inferior. On the other hand, it seems that in the case of the Scandinavians, they satisfied two very important criteria: they were blonde and had blue eyes!

One cannot speak about early Maltese migration to Australia without mentioning the notorious *Gange* incident of October 1916, which Dr. Barry York terms as '*one of the ugliest incidents in Australia's twentieth century immigration story*'. Perhaps, this event best illustrates the anti-Maltese sentiment that had been brewing in Australia over the years, fanned by the Australian media, which was also very hostile to the Maltese. '*The Brisbane Worker*' called them '*coloured job-jumpers*'. Journalist Donald Cameron was spreading the word that a Maltese invasion was imminent, while a speaker in an anti-conscription rally in Inverell announced that Australia had been invaded by 4,000 Maltese. This number was greatly exaggerated as, at the time, there were just over 900 Maltese in the whole country. As the war raged on in Europe, they were accused of taking the jobs of the young Australians who had taken up arms to fight for King and Country.

Their inadequate knowledge and grasp of English also did not help their cause as it hindered their adaptation into Australian society and way of life. The diocesan clergy in Sydney reported that the Maltese who established and integrated themselves successfully were only able to do so because of their knowledge of English. Reception and treatment of the Maltese in this city was particularly unpleasant. In fact, Maltese emigration to NSW was officially discouraged.

Few had heard about Malta, and Southern Europeans were not considered white. The Maltese raised suspicion as, although they called themselves British, they did not speak English or look Anglo-Saxon. Strickland always stressed that the Maltese were Christians and descended from the Phoenicians. However, the few Maltese in NSW gave their compatriots a bad name. They walked barefoot in groups of about thirty or forty, with one of them acting as their spokesman. They looked 'foreign', spoke no English, and instead, talked loudly in a strange language that sounded like Arabic.

In June 1912, a group of fifty Maltese landed in Sydney with no possessions. Only one of them could speak English. Australian workers refused to work with them and they immediately became a burden on the government. Cries of cheap and coloured labour and an *'influx of inferior races'* abounded, and Prime Minister Fisher and the NSW premier labelled them as *'undesirable migrants'*.

In September 1916, one month before the *Gange* episode, the ship *Arabia* arrived with ninety-seven Maltese aboard. They encountered resistance in the workplace as Australian workers refused to work by their side. A Maltese priest, Father William Bonnet, who had migrated earlier that same year, came to their rescue and managed to secure them jobs in Sydney and Mount Lyell. Father Bonnet always advocated the establishment of a Maltese community. When this true champion of the Maltese died in 1928, at the age of forty-four, he was deeply mourned.

The news and prospect of more Maltese arriving was not what the Australian Prime Minister Billy Hughes wanted to hear, especially so close to the conscription referendum in October. Amid rumours that shiploads of Maltese were going to invade Australia, taking the jobs of men who were away at war, he assured the Australian public that '*...during the War, no coloured labour will be admitted.*' His government stopped passports from being issued in Malta to migrants bound for Australia, and ordered new regulations prohibiting cheap labour. Undoubtedly, these measures were put into place to placate the labour movement that vehemently opposed conscription, sustaining its argument from the deep-grounded fear that the Australian workers serving the country would be replaced by cheap labour from overseas. This would have a drastic impact, resulting in the dwindling of hard-won employment standards.

The P&O Company was contacted to see whether any Maltese were on their way to Australia. The news was not good. 214 Maltese on the French steamer *Gange* were scheduled to arrive in Melbourne on their way to Sydney on the 28th of October, the eve of the referendum. The passengers on board had left for Australia before the passport restrictions came into force. The anti-conscription lobby seized this golden opportunity for their cause. The success of the referendum hung in the balance as Hughes's guarantee that no Maltese would be arriving was blown to pieces. The Governor-General cabled the Colonial Office in London that if the Maltese were allowed to arrive, this would *'kill the referendum'* and bring about a *'national disaster'*. He also requested the cooperation of the French government, since the vessel was French. The *Gange* had to be stopped from arriving in Melbourne before the referendum. Time was running out!

The *Gange*'s captain was ordered not to allow the Maltese passengers to disembark. Those on board were offered employment in munition factories in France, and alternative migration to Canada, New Zealand, Samoa, or South Africa. But they were adamant: they wanted to go either to Sydney or be sent back to Malta. But, it proved a difficult task to find a vessel in the middle of a war.

In the morning of the 29th of October, the *Gange* arrived in Williamstown. A Dutch professor went aboard and applied the dictation test in Dutch to the Maltese. Naturally, they all failed the test. The next day, the ship anchored at Port Melbourne and stayed there for three and a half days. The *Gange* then spent six days in Sydney. While there, forty-four managed to escape. Maltese wharf workers were suspected of helping them.

On the 12th of November, the *Gange* set sail for New Caledonia. In Noumea, they were well looked after, all expenses being covered by the Australian government. The Maltese were then taken back to Sydney in Berry's Bay, where they were detained in very bad conditions under armed guard on an old abandoned ship anchored at the harbour. In the meantime, they were receiving news of their starving families in Malta, who depended solely on the money their relatives overseas had promised to send them. More tried to escape.

Certain elements of the Australian populace were shocked at the treatment the Maltese were given. The Australian troops, returning from Gallipoli and the Front, were full of praise for the welcome and care they had received while convalescing in Malta. Everyone had heard about Malta's role in the care of wounded Anzacs, and yet the authorities had treated these Maltese so appallingly. They were pawns in the hands of both the Australian government and the labour movement, who used them to advance their respective agendas. A representative of the Maltese on the *Arabia* said: '*The reception to Australians in Malta was great, yet the Australians are rejecting us.*'

When news of the plight of the people on the *Gange* reached Malta, the Governor, Lord Metheun, expressed his concern that imperial unity was at stake. He was ashamed of the treatment the Maltese received and regarded it as a poor return for the kindness shown to the Anzacs in Malta. Metheun's concern was not purely selfless as he also had his own agenda: a stance motivated by the fear that the *Gange* incident would be used by the pro-Italian faction to fuel anti-British sentiment, and thwart his push to introduce English language education in Malta.

Sixty-five out of the 214 Maltese on the Gange had been at Gallipoli, including Gozitan Emmanuel Attard who had served with the Maltese Labour Corps. They had excellent references from the British Military Authorities about their service and they had also served at Lemnos and Mudros, but these accomplishments seemed to count for nothing. When interviewed by Dr. Barry York and Mark Caruana for their book *'Emmanuel Attard: From Gozo to Gallipoli'*, Emmanuel did not want to dwell on his war experience. He simply said: *'It is too sad for me'*.

The Australian Worker's Union was hard to impress. Trying to downplay and degrade Malta's efforts for the war, including its hospital role, it argued that the Maltese had not raised their own battalion. But this had not been done not by lack of will, but because the British War Office had not permitted it. Instead, they had to serve through other battalions or through the Maltese Labour Corps that provided more than 2,000 stevedores at Gallipoli and Salonika.

Throughout the war, 25,000 Maltese out of a population of 220,000 had volunteered. There were also eighty-seven Maltese officers, six of whom received distinctions for bravery. 650 Maltese served in the British Navy, and 300 joined the Canadian and Australian contingents. Only a few months before, in the battle of Jutland, various Maltese lost their lives on various battleships such as the *Queen Mary*, *Defence*, *Black Prince*, and the *Indefatigable*. Thus, although it was true that the Maltese did not fight under their own banner, their contribution to the war effort was not insignificant, as the A.W.U tried to purport.

On the other hand, support for the *Gange* passengers came from various sources. At the time, Gerald Strickland was the Governor of NSW but his powers were limited. He could only defend the victims. As he had done with the Maltese on board the *Arabia* a month earlier, Father Bonnet again did his utmost to help, rallying support also via the press. He brought the plight of the *Gange* Maltese to the attention of the authorities by writing to the Prime Minister and Governor General, committing himself that he would personally find work for those released. A number of pro-Empire groups and individuals, including returned servicemen, who had witnessed firsthand the kindness and generosity of the Maltese while convalescing in Malta, also offered their backing. It was pointed out that the destinies of both Malta and Australia were forged within the same Empire, and that the Maltese should not be considered as *'coloured labour'* since they are of European stock. Mrs. W.A. Walsh, whose son had died of wounds received at Gallipoli and was buried in Malta, demanded that the Maltese be allowed to land.

The Millions Club, a pro-immigration group led by real estate mogul Arthur Rickard, and '*The Sydney Morning Herald*' also called for the release of the men as the readers were assured that the Maltese were of Caucasian stock.

In March 1917, Prime Minister Hughes met with Father Bonnet and had a change of heart. The men were finally released on the 13th of March 1917, six whole months since the beginning of their ordeal. All were eventually placed in jobs and allowed to join unions. A number of them went to Tasmania to Mount Lyell Company, where a number of Maltese had been working since 1912. Others travelled to Queensland to work in the sugar cane fields. Those who could not find work were supported by Strickland with funds from his charity, while six others were repatriated to Malta on medical grounds, after being diagnosed with trachoma.

Three protagonists of the *Gange* saga who remained in Australia were brothers Tony and Joe Apap, and Celestino 'Charlie' Bajada. Tony ended up working in various jobs from Sydney to Queensland to Darwin. His brother started the Red Lea poultry business at Blacktown. By the late 1930's, Charlie Bajada, who had only been fourteen years of age when he accompanied his father on the *Gange*, was widely recognised as one of the top poultry processors in NSW, after establishing Baiada Poultry, which is still going strong today. Far from being a burden on Australia, the success and valid contribution by Tony, Joe and Charlie in their adoptive country proved all the detractors and naysayers wrong.

Following the *Gange* incident, Prime Minister Hughes became sarcastically known as William 'Maltese' Hughes and the *Gange* victims as *'the children of Billy Hughes.'* But, far from being the benevolent father, Hughes did not want any Maltese migrants at all. This might have been a question of sour grapes after his defeat in the referendum. Maltese immigration to Australia remained officially prohibited until 1920.

*25. New Year wishes from Charles*

Thus, in these early years of migration from Malta, the Maltese who wanted to seek a better life beyond the shores of their country were largely left to their own devices. They made huge sacrifices to raise the funds for their passage with no help from anybody, then they had to say goodbye to their loved ones, probably seeing them for the last time. After surviving the long and perilous sea journey, they arrived in a strange land where they had to do work they had never done before: on railway lines, in mines and vast sugar cane fields, using tools they had never laid eyes on. If a few had a university degree from Malta, this did not count for much as it was not recognised. Despite these difficulties, like most of the small Maltese migrant community at the time, Charles, Waldemar and Anthony survived and thrived in their new adoptive country. This was due solely to the support they gave each other within their tight knit community, and their unwavering faith, determination, and inner strength.

# CHAPTER III

# *In Love*

***'I love Carlo more than all the world, far, far more, and if God will spare him to us all I shall never cease to try to make him happy.'***
*(Letter from Gertie to Charles's family)*

After being in Australia for two and a half years, Charles, in a letter to his father, took stock of his life. He thanked God for what he had, but commented that earning £187 and 10 shillings a year was not enough to live comfortably, and enjoy life like a young man should. He told his father that he used to end up spending all his money as he enjoyed living life to the full.

Finding a secure job shortly after his arrival helped Charles to integrate easily with the local community. His jovial and pleasing manners also attracted numerous friends, and he was frequently the guest of various affluent families whose daughters he took out to dances, concerts, races, and picnics. We also find him active in the Cottesloe Surf Life Saving Club where, in March 1914, he was awarded a bronze medallion and proficiency certificate by the Royal Life-Saving Society.

In a letter to one of his sisters, Charles replied to her comment that he was considered quite a catch by all the girls. He admitted that he did have a lot of friends of the fairer sex, but not a special one. In a letter to his mother Giorgina, dated the 22nd of June 1912, in reply to her question about the girls he was friendly with, he confessed his fondness for flirting, but that this was something very much acceptable in Australia. There were also plenty of opportunities for socialising and he took full advantage of this. According to Charles, in Australia, which was a much more liberal society than Malta, girls and boys were very open with each other, even if they had just met. He stressed that the important thing was that everyone knew where to draw the line. On the subject of kissing, he said that: '*You can be sure that when I get the chance I am not found wanting. But I know the way they do it and I do the same, when in Rome do as the Romans do.*' Commenting on the girls he was meeting, he stated that some were more beautiful than others, but the latter may talk better than the former!

***26. Christmas greetings from Charles to his sisters***

Writing to his sister Melita, Charles gave her advice about love in relation to a man called Beppe, suggesting that she should show her true feelings to her love interest:

*'I have often thought, not to tell you that I experienced it, that when a person really loves another person, he or she can easily make love be returned by showing*

*him or her the true love. Do you show him how much you love him? What is the good of loving him if you don't show him such love? He would be gaining nothing while you would be losing the love which he would have returned had you shown him yours.'*

Despite this wise brotherly advice, Melita's relationship with Beppe did not have the desired fairytale ending, and she would later marry an American banker called Richard Rowe, and settle in New York.

In another letter to his sister Laura, dated the 23rd of May 1913, he mentioned a French girl called Camille whom he admired for her sense of Parisian style. But their relationship ended by mutual consent as people started to gossip behind their back, spreading nasty rumours. He also mentioned a girl called Gertie, whom he described as one of the best pianists in town. It seems though that Charles did not just admire her for her musical abilities as he also confessed to his sister that *'she and I are very friendly together'*. The relationship with Gertie progressed well beyond a platonic friendship. Charles's family in Malta were over the moon about his choice, singing Gertie's praise, especially when they received a beautiful photograph of her. Even his father liked what he saw:

*'We liked her very much. Mama, sisters and the few who have seen her picture are enthusiastic over her; I do greatly appreciate your choice.'*

When Charles sailed off to war, Gertie wrote to the Bonavia family in Malta, happily announcing that they had become engaged before he left Australia:

*'I love Carlo more than all the world, far, far more and if God will spare him to us all, I shall never cease to try to make him happy...... When you are praying for Carlo and feeling unhappy, will you give one thought to a lonely girl across the seas who loves him better than life?'*

With his demanding job and active social life, Charles sometimes found it hard to find the time to keep in touch with his relatives in Malta. Replying to his father's accusation that he was not writing often, Charles retorted that he doubted whether boys in his similar circumstances wrote as often as he did:

*'It is not carelessness as you say and it is not overwork, it is not pleasure seeking, nor even a 'bella donna' who is keeping me from writing often.'*

When his university mates, Waldemar and Anthony, arrived in Perth in

February 1912, he introduced them to his wide circle of friends and helped them to settle in. Like Charles, both Waldemar and Anthony were well-spoken and oozed charm. Without any doubt, soon after their arrival, with their foreign good looks, fashionable style, and exotic accents, they would also have become popular on Perth's social scene, especially among its female population. The three of them were active in the community and furthermore, they contributed to various charities such as the Silver Chain and the Belgian Nuns Fund.

Among the collection of old sepia photos belonging to the Beck family, a few portray a dark-haired Maltese beauty called Evelyn or Velina Calleja, who came from the town of Birkirkara. These photos are all signed at the back with a dedication: *'To Waldemar with fondest love, Evelyn'* and *'To dearest Waldemar from yours ever Evelyn 1 July 1915'*. Thus, Waldemar and Evelyn kept in contact long after he left Malta.

*27. Velina next to Waldemar before his departure*

In a letter to his sister-in-law Vira, dated the 14th of May 1916, Waldemar wrote:

*'Għażiża Vira,*

*Illum ġiet għall-ittri. Ktibt lil ħija, lil Ġiusi, lil Velina u issa lilek......'*

*('Dear Vira,*

*Today it is my letter-writing day. I have written to my brother, to Ġiusi* (nephew Josie) *to Velina and now I am writing to you.')*

In the same letter, he wrote that Velina had written to him, and that his letter to her had not addressed her concerns. He told Vira that although he understood she did not want to get involved, he could not have been clearer to Velina. The young lady had apparently received various proposals of marriage. Waldemar was feeling guilty that she seemed to have rejected her suitors because of him. He felt that, for the moment, the best thing to do was to give her the freedom to move on with her life, rather than wait for him and risk passing on a good offer of marriage:

*'Velina kitbitli. L-ittra tiegħi ma kinitx risposta. Naf li inti ma tridx tindaħal imma jien jidhirli li ma nistax niktbilha aħjar. Kellha żewġ mistoqsijiet*

*28. Velina*

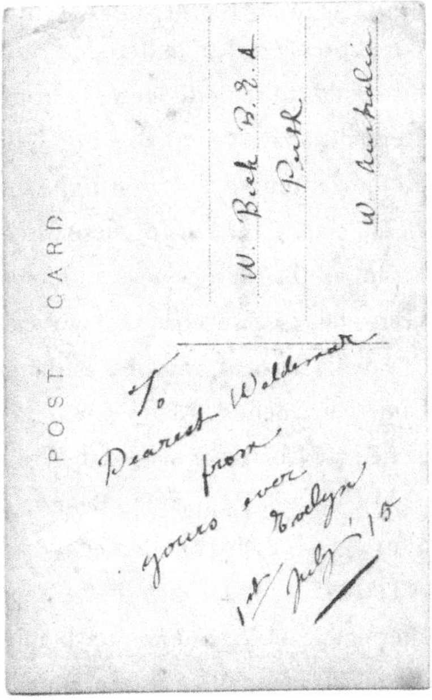


*29. Postcard from Velina to Waldemar*

*għaż-żwieġ u diġa' kellha oħrajn u jien ma nistax/rridx (?) niżżewiġha, jien inkun il-ħtija.*

*Għalhekk jidhirli li la il-proposti huma ħżiena, għalissa aħjar inħalliha libera. Jekk il-quddiem jameljoraw forsi nkun nista' niktbilha imma kif inhuma l-aħjar ħaġa hi li tkun libera.'*

In the same letter, Waldemar wrote that although he rarely mentioned the fact that he missed his family in Malta, he thought of them often. He had a picture of his mother on top of his bed but one time, the hooks came loose and it fell right on his head! Luckily, the glass did not break but he was frightened out of his wits and was also hurt on impact. He fixed it with a piece of wire:

*'Għad li ma tantx insemmi, jien niftakar fikom ilkoll. Ir-ritratt tal-mama qiegħed sewwa sewwa fuq rasi u ilu ftit il-... ittieklu u l-inkwatru waqa' fuq rasi. Il-ħġieġa ma inkisritx imma qbiżt qabża bil-qatgħa u bid-daqqa li ħadt. Issa għamiltlu biċċa wire.'*

Waldemar ended the letter by urging his sister-in-law to write to him often. These early letters are quite enjoyable to read, and make quite a contrast to the grim letters of a few months later when war broke out. Upon reading, one gets the impression that, although it must have been hard at first for Charles and his friends to settle so far away from their homeland and find good jobs, they seemed to have been living life to the full. Unfortunately, these light-hearted and jovial times were soon to be overshadowed by the clouds of war, which would change the lives of these three friends forever.

One of the reasons why Waldemar decided to let Velina loose could have been a tall, elegant Australian girl called Barbara Gugeri, who he met in Perth.

Amelia Barbara, known as Barbara, was the eldest daughter of Peter Anthony Gugeri and Belinda Mary Morris. Peter had been previously married to Catherine Jane Jackson, who had died on the 17th of April 1889, after bearing him eight children. He married Belinda on the 27th of October 1890. The second marriage produced another eight children.

Leo Van de Pas, a friend of a Gugeri relative, researched the family history after interviewing numerous family members and studying Peter Gugeri's diaries. Peter's father was originally from Switzerland and had moved to England where Peter was born. A crucifix and a porcelain figurine that had

*30. Barbara*

been brought to the new home by the family were donated to the Benedictine monks of New Norcia. The founder of the only monastic town in Australia, Bishop Rosendo Salvado, was a great friend of Peter Gugeri, whose portrait still exists at the monastery to this day.

Peter had settled in Western Australia in 1871, where he established a vineyard in the Swan Valley. In 1874, he married his first wife. Later, he became involved in various businesses and public offices. His last appointment before retiring in 1910 was as Clerk of the Courts at Midland. His second wife Belinda survived him by fifteen years, dying in 1945.

Barbara had an eventful start to her life. She was born at Hay Street West in Perth on the 27th of December 1891. When she was only fourteen days old, her mother became ill with typhoid fever for nearly six weeks. Barbara was then handed to the care of Mrs. Hardwick of West Perth.

As soon as Barbara's mother got over the fever, Barbara was brought back to the vineyard after spending the first night at the Royal Hotel. But she cried so much, they did not know what to do, so they took her back to Mrs. Hardwick for the night, and Barbara was returned to her mother the following day. She had been very sick in Perth because of the milk there but back home, her health started to improve.

When I contacted Barbara's family in Australia, they all were aware of the Maltese connection through Waldemar in their aunt's life. At the home of her niece, the daughter of Barbara's brother, Leo, I went through a number of family albums which contained photos of Waldemar and his family in Malta, and a postcard from Waldemar's nephew Josie to his uncle telling him that he missed him and wanted to see him *'very much'*. When a few months later, I met

Waldemar's family in Malta, it was amazing to see copies of the same photos in their family collection!

*'At home I could not believe that I actually met you and I kept thinking about what you told me about your auntie and kept running through my mind the images on the photos.'* (Email I sent to Barbara after meeting her in 2001)

*'I still find it hard to believe that after all this time of knowing that there was a person called Waldemar Beck associated with our family, that suddenly he has become so much more of a personality and that there is a family who, thanks to your research, have become real. Very exciting!'* (Email from Barbara, 11th of October 2001)

*'I wonder if Mariella's father knew anything about Aunt Barbs – I do hope so because that would bring us all together a little bit more - it is all very interesting and really quite amazing because only a little while ago he (Waldemar) was Aunt Barbara's fiancé and a face in some photos but we knew not much more about him.'* (Email from Barbara, 18th of October 2001)

Barbara's sister Debbie, who lives in Western Australia, was also '*fascinated to hear all about your involvement with his family as I have always known of him.*'

Besides a couple of photos, Debbie also has a silver pendant that used to belong to Waldemar with an inscription at the back:

'No.6969 Pte W. Beck 23/16 A.I.F'

*'It almost feels like a part of the family has been found which is terribly exciting... What a good looking man he was too. I love family history and to think you are actually in contact with his family is wonderful and he is now not such an unknown, until now only heard of, entity.'*

(Email from Debbie Gugeri, 11th of October 2001)

*'Thank you so much for all the information about Waldemar and his family and our family. It is all so exciting. I do wish I was there with you to meet them all.'*

(Letter from Barbara Gugeri, 8th of January 2002)

Both Barbara and Debbie have very fond memories of their aunt Barbara and her sisters Dorothy and Deborah, who never married. The three sisters were described as fiercely protective and proud of the Gugeri name.

*'I remember Barbara as being absolutely delightful and terribly elegant. Deborah was more austere and I think would have been the one who made the*

*31. Silver locket*

*decisions and had the final say in most matters. The far more dominant one. The last sister alive was Dorothy. I spent quite a lot of time visiting her at St John of God hospital in Perth where she spent her last years and far more of them than she expected to. She always related fascinating stories about their family growing up, always referring to their father as 'Daddy' which I thought was wonderful, her being in her nineties at the time. In fact they all did, now I come to think of it.'* (Email from Debbie Gugeri, 17th of October 2001)

Yvonne Kelly, a granddaughter of Peter Gugeri, recalls:

*'I lived for seven years with her, you know with Aunt Dossy, Barbara and Deborah in Mount Street. Dossy used to live in the hospital as a matron, but she used to come home with us on her days off. Aunt Deborah was in charge of the first floor in the Mount. So two nurses in the house! You can imagine! Poor Aunt Barbara was the meat in the sandwich and I was a bit dragged along at the same time, because I sort of kept house for them. We had the shop at the same time.'*

In his recollections dated the 24th of November 1986, Gerald Marcus Glaskin, Peter Gugeri's great grandson, records that:

***32. Barbara with her mother and sisters***

*'Aunt Dossy seemed somehow remote and less demonstrative than Aunt Barbara, even when she became more familiar to me as a young man and she became a nurse and later matron with the Rehabilitation Department, caring for returned veterans.'*

Waldemar's great-nephew recalls that there used to be a *'bagoll'*, a big, heavy suitcase, which used to belong to *'Ziju Volly'* (Uncle Volly or Waldemar). This had been sent from the War Office and included, amongst other things, a penknife, which later Vira would use to repair Waldemar's sandals. The rosary beads forming part of Waldemar's kit seemed to have been made of the seeds of some Australian native tree.

Waldemar had sent his family Aboriginal weapons and other native artefacts, including two walking sticks and a wooden snake, which had to be cut in half in order to be posted.

There was also a ring with an inscription inside, *'Love Barbara'*, which Waldemar's great-nephew used to wear; but this item has been lost over the years. Even the things in the suitcase have all been thrown away as most of them had become rusty and old.

We can assume that Waldemar and Charles's friend, Anthony, also did not lack female attention. A reference to Anthony's charming personality was made by Deric Thompson, who worked with him in Darwin. In an interview recorded in the 1980's, which is preserved at the Northern Territory Archives, Thompson describes Anthony:

*'Tony was fluent in Italian and French and he was quite a man about town. He was short, but more personality in his little thumb than I would have in three bodies.'*

Thompson also alludes to the fact that Anthony was quite popular with the opposite sex:

*'But he was a great ladies' man, great conquistador, I should think. And he was a great favourite with any person who came to Darwin, because he was entertaining. And he was always sought-after because of his knowledge of languages and his wide range of reading.'*

He mentions that Anthony was given the task by Mrs. Abbott, the wife of the Northern Territory's administrator, to teach her daughters French:

*'...but I don't ever think they were left alone, nor do I think they ever ...became great French students.'*

Peter Castaldi also refers to Anthony's charismatic personality:

*'The photo is a match for the one I have, he is in WWI uniform. He was known as Tony and was charming. He called my father Ġiġi.'* (Email dated 16th of November 2009)

The first few months of their settlement in Australia would not have been easy for Charles, Waldemar and Anthony. Far away from their families, with their unrecognised university qualifications, at first they would have struggled to secure a good job which would ensure them a decent life. As they finally started to enjoy their youth, like any other normal young men in their twenties, making friends and discovering love, the world around them was changing rapidly. Far away in Europe, dark clouds were gathering and the world started to brace itself for the inevitable. Their carefree days would soon be over.

## CHAPTER IV

# *The Winds of War*

***'...I am afraid that only a miracle from God can save Europe now.'***
*(Letter from Charles to his family, 2nd of August 1914)*

Kaiser Wilhelm II of Germany, one of the major players in the First World War, was rarely known for his diplomatic nature. However, years earlier, in 1900, he had lent his support to Archduke Franz Ferdinand, heir to the Austrian throne, in his desire to marry commoner Sophie Chotek, against the wishes of his uncle, Emperor Franz Joseph. Together with Pope Leo XIII and Tsar Nicholas II of Russia, Wilhelm interceded on Franz Ferdinand's behalf to the Emperor, arguing that the disagreement was undermining the stability of the monarchy. The Emperor, seeing that his nephew refused to consider anyone else, bowed to the pressure and agreed to the marriage. Despite this bumpy start, the union between Franz Ferdinand and Sophie seems to have been a happy one. How unfortunate that it ended in a tragic way with global repercussions.

On the 28th of June 1914, Franz Ferdinand, together with his wife, was on an official visit to Sarajevo, the capital of Bosnia-Herzegovina. Until 1908, Bosnia, considered to be the heartland of Southern Slavs, had been part of the Ottoman Empire. The Archduke had been warned of the danger of his visit due to mounting pressure from Serbian nationalists who were urging Bosnia to leave Austria and join its big brother, Serbia. Things were also looking pretty grim for the future of the Austro-Hungarian Empire as it was steadily on the decline. However, few would have predicted that, what should have been a routine visit by a representative of the Austrian Emperor, would end up producing the single event that triggered a war that, would not only change the world, geographically and politically, forever, but also have tragic, long-lasting effects on a whole generation.

The Balkans had been a boiling pot for quite a while with Russia, Austria, and Turkey on a continuous clash course. Russia was scared that Austria would gain control of Constantinople and thus block Russia's route to the Mediterranean. Two wars had already been raging in the Balkans between 1912 and 1913. These conflicts revolved around the redistribution of Turkish-controlled territories. Serbia, supported by Russia, yearned for independence from Austria and stood at the forefront in the quest to unite all Slavs. This Serbian ambition was one of the sparks that started a war that had been brewing for a long time.

When the first day of the 20th century dawned, the world was bursting with optimism and hope as its future looked rosy. Breakthroughs and progress were making life easier and more comfortable. The undisputed super powers of this era were Britain, France, and Germany who competed heavily against each other, especially in the colonial race, although Germany lagged far behind in territorial supremacy. The United States, whose population was fast increasing through immigration, was also slowly but surely emerging as a power to be reckoned with.

In these early years of the 20th century, the world resembled a chess board, where the major political players all knew their place. Any change of position was calculated and meticulous, and if another country had its rights infringed, allies would swiftly come to each other's aid for protection. Politically, a number of agreements between allied countries kept conflict at bay and maintained a fragile peace.

In 1914, the Triple Alliance consisted of Italy, Germany, and the Austro-Hungarian Empire. A secret treaty between these three countries had been signed in 1882 to counter France's ambitions towards Tunisia. The signatories pledged that if one of them was attacked, the others would either remain neutral or defend their ally in trouble.

On the other side, France and Russia had forged their alliance since 1894. Ten years later, in 1904, Britain and France signed the '*Entente Cordiale*' and, three years later, were joined by Russia, each guaranteeing to come to the aid of the others in the event of war. Belgium remained a neutral country and, in an 1839 treaty, Britain had pledged to defend Belgium's neutrality. In 1912, Britain and France further strengthened their cooperation by deciding that Britain would have the responsibility of patrolling the North Sea, while the French would monitor the Mediterranean. Thus, what started out as a friendly agreement had developed into a military alliance.

The world stage was further complicated as the major powers were linked to each other by marriage. Kaiser Wilhelm was the cousin of the British King George V, both having Queen Victoria as their grandmother. The Kaiser was also the cousin of the Russian Tsarina Alexandra. In January 1913, Wilhelm wrote a very cordial message to her husband, Tsar Nicholas II:

*'I sincerely hope and believe that 1913 will flow peacefully as you telegraphed me on New Year's Day…I think that we can both look on the future quietly.'*

Wilhelm's wishes for a peaceful future proved to be very ironic in the light of developments in the months to come.

At the outbreak of hostilities, Tsar Nicholas telegraphed the Kaiser:

*'With the aid of God, it must be possible for our long-tried friendship to prevent the shedding of blood.'*

Tsar Nicholas was also a cousin of King George as their Danish mothers, Maria Feodorovna and Queen Alexandra, were sisters, and to add to the mix, his mother-in-law was Alice, one of Queen Victoria's daughters! However, as the war unfolded, there was not much love lost between the fighting factions, and these family ties were quickly forgotten.

Ironically, Kaiser Wilhelm looked up to Britain, and wanted to fashion Germany on his cousin's country. In August 1889, upon his return home from a visit to England to see his grandmother Queen Victoria, Wilhelm thanked her for her hospitality, adding that he would start taking an interest in her fleet '*as if it were my own, and with keenest sympathy shall I watch every phase of its further development.*' During his visit, Wilhelm had inspected English naval equipment and declared that one day the German fleet would surpass the British one. Twenty years later, in 1909, the Kaiser visited Malta, where, undoubtedly, he would have admired the hustle and bustle of the picturesque Grand Harbour dotted with British ships.

Germany's ambition leaned more towards becoming the dominant power in Europe, rather than to conquer far-away countries and establish itself as a colonial power. Since its unification in 1871, when it had largely been an agricultural nation, Germany had made great strides in industry and technology, and in just thirty years became the third most industrialised country after the United States and Britain. By 1914, this nation had become the leading producer of iron, steel, coal, and cars. This was indeed a golden era for Germany, with its industry booming, and Berlin fast becoming the centre of knowledge. On the other hand, Britain had established itself as a strong imperial nation with a long tradition of victories and success in combat. In 1906, Britain launched the new Dreadnought battleships, which were much quicker and could carry heavier guns. The British Navy could also boast of nearly eighty destroyers, ten armoured cruisers and another ten battle cruisers, twelve light cruisers and more than twenty battleships, among them the *Revenge*, *Iron Duke*, and *Hercules*. In the face of these ever-growing British naval ambitions, Germany could not be left behind. The arms race with Britain was well under way. Further east, in Russia, the beat of the revolutionary drum had started to gather momentum, with Tsar Nicholas considered a weak leader, despite having the largest army in Europe.

Amidst these underlying currents of tension between the major power players and continental instability, everyone was waiting with bated breath for a spark to ignite the whole of Europe: no one would have guessed that this would occur in Sarajevo by a nineteen-year old Serb nationalist, Gavrilo Princip, on the fateful morning of the 28th of June 1914. Incidentally, this was already a significant date in the Serbian calendar as five centuries earlier on the day, a famous battle had been fought between the Serbs and the Turks.

Princip was a member of the terrorist group Black Hand, a nationalist movement favouring a union between Bosnia and Serbia. The two shots he fired would trigger the bloodiest war in the history of mankind: the first bullet struck the Archduke's throat, the second hit Sophie's stomach. With blood pouring from his mouth, Franz Ferdinand implored his wife *'Sophie, Sophie don't die. Keep alive for our children.'* It was too late - Sophie was already dead and Franz Ferdinand breathed his last by the time the car reached the Governor's residence. In May 1918, just a few months before the end of the war, Princip died in prison from tuberculosis. He lived long enough to see the devastation his actions had produced. When asked whether he felt any guilt for precipitating a world war, he simply replied: *'If I hadn't done it, the Germans would have found another excuse.'*

The European nations responded quickly to this crisis and, like pieces on a chess board, all took their sides and positions. On one side, Britain led the Allies, which comprised France, Russia, Serbia, Montenegro, Belgium, Romania, Portugal, and Greece. They were later joined by Italy in 1915, and the United States in 1917. Italy had initially stayed out of the war, arguing that, in line with the Triple Alliance's 1882 treaty, it would remain neutral as neither Germany nor Austria were attacked. On the other side, Germany led Austria-Hungary, Bulgaria, and Turkey. At other times, Britain used to be the protector of the Ottoman Empire against the Russian Mediterranean expansion. However, as it had done during the Greek war of independence, in this instance it sided against the Turks, as it had started to notice increased close ties between Turkey and Germany. In fact, England decided not to hand over two war ships that Turkey had just purchased from the British navy. Quite understandably, Turkey was rather upset about this state of affairs and switched its allegiance to Germany.

On the 30th of July, the Austro-Hungarian Empire declared war on Serbia. Russia, regarding itself as the protector of the Slavs, started to mobilise its army of up to five million soldiers to intimidate Austria, rather than to attack it. Germany then issued a warning to Russia, and followed this up on the 1st of August with a declaration of war. It was then France's turn to get involved to help its Russian ally. Despite its huge force, in August 1914, the Russian army would suffer a huge blow at the hands of the Germans in Tannenberg, one of the very first battles of the war.

At the outbreak of war, Britain had 247,432 regulars and 218,280 reservists, France had 3.68 million, while Germany's army numbered 840,000. Belgium only had a small and inexperienced army.

The German plan was to first knock France out by invading it from the north, that is, via a free passage through neutral Belgium, where the French defences were weak. On the other hand, France was expecting an attack from the Swiss border. The plan was concocted by the German War Minister, Count Alfred von Schlieffen, and backed up by Chief of Staff von Moltke who sent over one million soldiers to march into Belgium on the 2nd of August. The next phase of the plan envisaged Germany then turning its full attention on Russia. The Kaiser boasted that *'In 15 days we will be in Paris.'* He even summarised the German plan as *'Lunch in Paris, dinner in Saint Petersburg.'* This plan was basically flawed as it wrongly assumed that France was weak, and Russia would take weeks to mobilise its troops.

Britain, respecting the informal agreement it had with France to support it if attacked, as well as its pledge to defend Belgium, sent an ultimatum to Germany to withdraw its troops from Belgium by midnight of the 4th of August 1914. When the Germans failed to do so, Britain announced that the two countries were at war. Germany had declared war on France the day before.

For the French, the war provided the perfect opportunity to pay back the Germans and seek revenge for the loss of the eastern province of Alsace-Lorraine to German control during the Franco-Prussian War of 1870-71. The resentment and bitterness after this loss were felt long afterwards. The French had high hopes that this coveted province would soon be flying the blue, white, and red flag again and that at long last, they would be able to heal their wounded Gallic pride.

In the meantime, on the other side of the world, Australia was in the throes of a federal election campaign with Prime Minister Joseph Cook battling against the Leader of the Opposition, Andrew Fisher.

When the British Empire made the call to arms, it was answered across the four corners of the globe: from the pungent forests and lakes of Canada to the spicy, exotic cities of India; from the dusty outback of Australia to the snow-peaked mountains of New Zealand.

At that time, Great Britain acted for all the Dominions in the conduct of foreign affairs, and most Australians considered themselves automatically at war with Germany, alongside Britain. On the 31st of July, Fisher, who would go on to win the election and become Prime Minister in September, uttered the famous words: '*Australia will stand by the mother country to our last man and our last shilling.*' Not to be outdone by his political rival, on the very next night, Cook declared: '*When the Empire is at war, so is Australia.*' In the fervent negotiations that preceded the war, the Australian government had already offered military support to strengthen Britain's hand. This offer followed the precedent set by the colony, prior to Federation, in assisting Britain in the Sudan, South Africa, and China.

Two days before the war was declared, on the 2nd of August 1914, Charles wrote to his family saying that he was missing the excitement of preparation for war in Malta. He said that Australia was too far away and '*although it concerns this country as much as others, people do not seem to realise the difficulty of the situation......... I am afraid that only a miracle from God can save Europe now.*' At the same time, Charles could see that Malta might reap some benefits from the dire situation '*by some trade during the preparations and during the war if it takes place, and if it does not, by the strengthening of the garrison and the Navy which are not over numerous for the present.*' He also reiterated his faith in the British Navy's power: '*Nothing can happen during the life of the British navy...... The great moment has come.*'

If Charles thought that Australia would not get involved due to its geographical isolation from the rest of the world, he was mistaken. On the 3rd of August, Prime Minister Cook pledged an expeditionary force of 20,000, and the ships of the Royal Australian Navy under the control of the British Navy in the event of war. As Minister for Defence, in 1909, Cook had established a military college and compulsory military training for boys aged between 12 and 26 years. This scheme, however, had been widely criticised as being an invasion of civil and religious freedoms. By 1914, about 210,000 teenagers were already being trained by Defence Department instructors. Cook had also concluded the agreement with Britain to establish the Royal Australian Navy.

When, a day after Australia's pledge, Britain declared war on Germany, jubilant crowds gathered in the major Australian cities. They cheered, celebrated, and sang *'Rule Britannia'* with heartfelt patriotism. Australia had only recently become united as a Federation but the ties with the mother country remained strong. Australians were fervent believers in God, the King, and the Empire and still viewed Britain as the motherland. If Britain was attacked, it would be their duty to answer the call and help Britain defend herself. A popular song of the time, *'Australia will be there'*, embodied this spirit and belief.

Straight away, men started to line up outside military headquarters in the big cities. Crowds gathered outside newspaper offices waiting excitedly for the special editions. Official recruitment for the Australian Expeditionary Force started on the 11th of August.

The call to join up was answered by men from all walks of life: clerks, clergymen, teachers, students, and bush men. Most ethnic backgrounds were also represented: Indigenous men, Anglo-Saxons, and migrants, including a number of Maltese. On the 7th of October 1914, *'The Daily Malta Chronicle'* published its first reports of Maltese volunteers in Australia. Seven of these would end up paying the ultimate price, while thirty others who survived the war returned to Australia. More than 200 Maltese living in Canada joined the First Canadian Contingent. One of them, Manuel Micallef from the town of Qormi, who had migrated to Canada, was killed in Belgium while standing guard on the 28th June 1915. A number of Maltese living in Tunis, Algiers, and France enlisted with the French army. A few formed part of the French fleet, which arrived in Malta on the 12th of August 1914. They were welcomed with throngs of people lining the bastions shouting *'Vive la France'*. Valletta and Floriana were teeming with French soldiers. It is one of the ironies of history that while the Maltese had fought hard to get rid of the French occupation between 1798 and 1800, 114 years later, the French fleet was welcomed with open arms!

In 1914, Malta's population numbered around 220,000. 25,000 of them volunteered for war-related service, including around 2,000 as part of the Maltese Labour Corps who would work side by side with the Anzacs in

Gallipoli, Lemnos, and Mudros.

When news of Britain's declaration of war against Germany reached Maltese shores, the local populace sang *'God save the King'* with great gusto and clapped cheerfully. Like the rest of Europe, they had been waiting for this moment and had long been mobilising in preparation for war. The Maltese expected the British navy to destroy the German navy in no time, and did not think there would be a ground war. Since Italy at first remained neutral, the Maltese also put their mind at rest that the war would not be too close. In fact, compared, say, to France and Belgium, Malta did not suffer much during this war. The situation would be completely different during the Second World War, when Malta earned the reputation of being the most heavily-bombed place on Earth!

As soon as Britain entered the war, Malta was put under martial law. Nets were placed around the Grand and Marsamxett Harbours, and restrictions were placed on the movements of boats. Houses had to be blacked out and all mail, books and newspapers were censored. Even the local newspapers had to pass through censorship before being sent to print. They could no longer publish notifications of arrival and departure of vessels. One printer got into trouble by publicising a soccer game and naming the ship whose crew was playing. The names of ships could not be printed or displayed in any way, and sailors stopped wearing their ship name on their cap, this being replaced with *'H.M.S'*.

The Maltese suffered from high prices for such products as cheese and eggs. Matches and kerosene also became scarce. This was mainly due to the presence of German and Austrian submarines that were hitting the supply ships in the Mediterranean. If you had a baby, or sick relative, you needed a doctor's certificate to get flour or semolina. Meat had also become a rare commodity and special permits were issued to sell horse and donkey meat. Bread had to be mixed with rice and corn, and caged birds were set free as there was no birdseed available.

Even entertainment was cut down. There were no lights or fireworks during the traditional feasts. Streets were only decorated with flags. The pealing of bells also had to stop as the sound interfered with the wireless. Later on in the war, this ban was lifted slightly and the bells could be rung on certain occasions such as feast days.

Back in Australia, volunteers from the country left their land and travelled by train, horse and cart, and even on foot, to the enlistment offices in the major cities. Groups marched long distances, collecting money and urging men to join up. Girls offered a kiss to those eligible to enlist as an incentive, and women were urged to send their men. Those who joined up gained the instant respect of society. Girls smiled readily at them and bus drivers let them get on for free.

The requirements for enlistment were: age between 19 (later reduced to 18) and 38 years (this was later increased to 45), height of 5 feet 6 inches (168 cm) with a 34 inch (86 cm) chest. In July 1915, the height was reduced to 5 ft 2 inches (157 cm). Two years later, it would be decreased by a further two inches to just 5 feet (150 cm).

Charles was one of the first who answered the call. According to records held at the National Archives in Canberra, he enlisted on the 24th of August 1914 at Helena Vale. He was single, 26 years and 11 months old, 5 feet 10 inches tall (178 cm), and weighed 142 lbs (64 kg). His chest measurement just met the requirement (34.5 inches). He was described as having a dark complexion, with dark brown hair and eyes, and one scar on his right knee. His *'trade or calling'* was that of a draughtsman, and his *'religious denomination'* Roman Catholic.

On the 9th of September, Charles was declared fit for service at Blackboy Hill by the Medical Officer, Captain E.T. Brennan. Recruits were examined and tested for a number of conditions, ranging from *'defective intelligence'* to hernia, varicose veins, contracted and deformed chest, syphilis, scrofula, abnormal curvature of the spine; *'or any other disease or physical defect calculated to unfit him for the duties of a soldier.'* Charles was also considered to see *'the required distance with either eye; his heart and lungs are healthy; he has the free use of his joints and limbs; and he declares he is not subject to fits of any description.'* On the same day, he took the oath in front of the Attesting Officer Major E. Drake Brockman:

*'I, Charles Emanuel Bonavia, swear that I will well and truly serve our*

*Sovereign Lord the King in the Australian Imperial Force from 9/9/14 until the end of the War, and a further period of four months thereafter unless lawfully discharged, dismissed, or removed therefrom; and that I will resist His Majesty's enemies and cause his Majesty's peace to be kept and maintained; and that I will in all matters appertaining to my service, faithfully discharge my duty according to law.'*

Following these formalities, Charles became Private 157 with the 11th Battalion, 3rd Infantry Brigade, nominating his father as his next of kin. His daily rate of pay was five shillings.

After enlisting, hoping that he would be sailing to Europe soon, Charles wrote to his sister, Laura:

*'Give a dear nice fond kiss to Mama, tell her that you dreamt that I came to you and told you to give a kiss for me. Do that and ask her if she will be pleased when I will be with you again.'*

Anthony enlisted on the 26th April 1915, one day after the Gallipoli landing. 'The *Daily Malta Chronicle*' announced his enlistment on the 14th of August 1915:

*'The number of Maltese serving with the Forces of the overseas Dominions is swelling daily. Hardly a week elapses without our hearing of new enlistments of countrymen of ours in the ranks of the Australasian or Canadian Forces. The latest enlistments include, according to letters from*

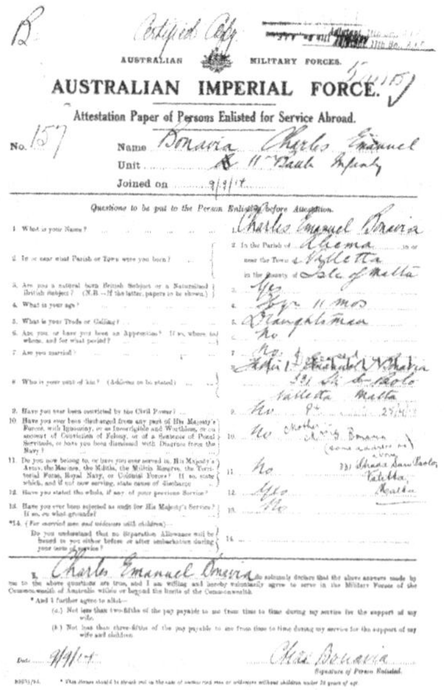

Certified Copy

AUSTRALIAN MILITARY FORCES.

AUSTRALIAN IMPERIAL FORCE.

Attestation Paper of Persons Enlisted for Service Abroad.

No. 157 Name Bonavia Charles Emanuel

Unit 11th Batt. Infantry

Joined on 9/9/14

Questions to be put to the Person Enlisting before Attestation.

1. What is your Name? — Charles Emanuel Bonavia
2. In or near what Parish or Town were you born? — In the Parish of ... near the Town of Valletta in the County of Isle of Malta
3. Are you a natural born British Subject or a Naturalised British Subject? (N.B.—If the latter, papers to be shown.) — Yes
4. What is your age? — 11 mos
5. What is your Trade or Calling? — Draughtsman
6. Are you, or have you been an Apprentice? If so, where, to whom, and for what period? — No
7. Are you married? — No
8. Who is your next of kin? (Address to be stated) — Valletta Malta
9. Have you ever been convicted by the Civil Power? — No
10. Have you ever been discharged from any part of His Majesty's Forces, with Ignominy, or as Incorrigible and Worthless, or on account of Conviction of Felony, or of a Sentence of Penal Servitude, or have you been dismissed with Disgrace from the Navy? — No
11. Do you now belong to, or have you ever served in, His Majesty's Army, the Marines, the Militia, the Militia Reserve, the Territorial Force, Royal Navy, or Colonial Forces? If so, state which, and if not now serving, state cause of discharge — No
12. Have you stated the whole, if any, of your previous Service? — Yes
13. Have you ever been rejected as unfit for His Majesty's Service? If so, on what grounds? — No
14. (For married men and widowers with children)— Do you understand that no Separation Allowance will be issued to you either before or after embarkation during your term of service?

I, Charles Emanuel Bonavia do solemnly declare that the above answers made by me to the above questions are true, and I am willing and hereby voluntarily agree to serve in the Military Forces of the Commonwealth of Australia within or beyond the limits of the Commonwealth.

* And I further agree to allot—

(a.) Not less than two-fifths of the pay payable to me from time to time during my service for the support of my wife.

(b.) Not less than three-fifths of the pay payable to me from time to time during my service for the support of my wife and children.

Date 9/9/14 Chas Bonavia Signature of Person Enlisted.

* This clause should be struck out in the case of unmarried men or widowers without children under 18 years of age.

33. *Charles's enlistment papers*

THE

Daily Malta Chronicle

SATURDAY, 14TH AUGUST, 1915.

SUNSET, TO-DAY ... 6 hr. 55. min.

THE EUROPEAN CONFLICT.

34. *The Daily Malta Chronicle edition announcing Anthony's enlistment*

*Australia, Mr A. Xuereb, L.S & A, the son of Dr F. Xuereb, who proceeded to Australia some years ago.'*

Enlistment from Darwin, where Anthony was living at the time, was not encouraged by the authorities as they were scared that if too many enlisted, no one would be left behind to protect what was considered a strategic outpost. Another less honourable reason was the fear that if many white men enlisted, Darwin would be overrun by ethnics. The Minister of Defence wrote to the Administrator that it was not envisaged that a voluntary corps would be raised from remote areas. When men from Darwin insisted they wanted to contribute directly to the war effort, they were told they could travel to Queensland to enlist at their own expense. A number of them did so, travelling to Townsville. By the middle of March, the military bigwigs had a change of heart that allowed the Northern Territorians to have their own battalion. Anthony formed part of this first contingent who left on the ship *Changsha* on their way to Enoggera Barracks in Brisbane. On the day of their departure they were feted and farewelled at the Darwin Town Hall, and accompanied by the local brass band to the wharf.

As for Waldemar, he would enlist in September 1916. In a letter to his brother George sent in October 1914, Waldemar expressed his concern about the effects of war on Malta and his family:

*'I am A1 and hope all of you are the same. I can tell you, you are much more in my thoughts now than ever before though I sincerely hope that Malta will come to no harm whatsoever. As regards out here of course we are allright. Give my love to all.'*

By the end of 1914, 52,561 Australian men had enlisted. At the time, the population of Australia was just under five million. Enthused with romantic patriotism, under-age boys lied about their age to army recruiting officers. This was made easier since, upon enlistment, they were not asked their date of birth, but only the stated age. As the war wore on, though, and morale began to plummet, the number of volunteers started to dwindle, and it was not long before the debate on whether to introduce conscription or not started to rage. More than 400,000 Australians would enlist throughout the war; over half of them would either be killed or injured.

Each battalion of the A.I.F was recruited on a broadly regional basis from the six states of Australia. The infantry consisted of five divisions, each made up of three brigades. In turn, each brigade contained four battalions. The ideal number of soldiers in a battalion, the main frontline infantry fighting unit, is a 1,000, but by the middle of 1918, due to the heavy casualties and reduced number of reinforcements, this target was rarely reached.

Furthermore, one would find also the corps troops, which consisted of support units such as clerks, orderlies, draughtsmen, grooms, cooks, and servants. The Mechanical Transport Units maintained trucks for distribution of food, ammunition, and stores of all kinds, while other vehicles were maintained by the Mobile Ordinance Workshops. The Australian Army Medical Corps took care of the casualty clearing stations and field ambulances. Other sections included the Australian Cyclist Battalion and the Signal Troop that used telegraph, telephone, wireless, aeroplanes, dispatch riders, dogs, and even pigeons to get messages across. The Number 3 Squadron, made up of the Australian Flying Corps and the Captive Balloon Service, was the eyes of the Corps, informing ground troops of enemy movements. The Australian Light Horse Brigade, which would make such a name for itself in the deserts of North Africa, sent 160,000 Australian horses overseas during the war, of which only one returned!

In January 1915, the eight companies of the A.I.F battalions became four and in 1916, the Anzac troops reorganised into two corps: I Anzac corps made up of the 1st and 2nd Australian and New Zealand divisions, under the command of General Birdwood, and II Anzac Corps incorporating the 4th and 5th Australian Division under General Godley. The I Anzac Corps joined Kitchener's New Army in France under General Sir Douglas Haig and was deployed near Armentières in the northern sector of the Western Front. ANZAC stands for the '*Australian and New Zealand Army Corps*', and the term 'Anzacs' refers to the men forming part of the corps. The term started to be used in this sense at Gallipoli.

The reasons for enlistment were various. In 1914, a severe drought had ravaged the countryside which reduced the wheat crop by two thirds. A large number of the volunteers were unemployed and saw a stint in the military as a short term solution to their financial difficulties.

*'...But the hard times are being felt over out here. There is such a lot of unemployment. Most of the farmers have had to see out owing to a bad season so all the farm hands in those places have come to Perth. In fact work is so scarce that people are accepting jobs at 5 shillings a week. I can tell you that I count myself lucky just at present to have a job of any sort.'*

(Letter from Waldemar to his brother George, 2nd of October 1914)

Another main reason for enlisting was a sense of duty and fervent patriotism. Other young men enlisted to satisfy their thirst for adventure. Others had no choice but to bow to pressure from wives, girlfriends, fathers, or friends. Some saw enlistment as the perfect opportunity to prove their manhood or get away from family problems, such as a loveless marriage, or joined up to be with their mates. Others went so far as to enlist for the sole purpose of avoiding the abuse hurled by elderly women at those who had not enlisted without having a good reason. Men who had been rejected due to health or other reasons were given a special badge so they would not be mistaken for the *'shirkers'* or *'cold footers'*, who were given a white feather by militant ladies who shouted *'Join the Army'* as they pinned the feather unceremoniously on their lapel. Posters were put up around towns and cities depicting a young man idling his time away, while an onlooking soldier asks: *'Which picture would your father like to show his friends?'*

A large number of volunteers also wanted to express their strong animosity towards the Germans. The Allies' propaganda machine depicted the Germans as inhuman brutes, and likened the Kaiser to the ruthless warrior Attila the Hun or the antichrist. Images of Saint George slaying a German dragon were circulated. Rumours of German atrocities in Belgium were rampant; for instance, it was widely believed in Britain that the German armies in Belgium were bayoneting babies. These continued to fuel the mounting anti-German sentiment and made civilians believe in the just cause of the war, instilling in everyone a desire to do their bit against the Teutonic Hun. However, Vera Brittain, a voluntary nurse who was posted for some time in Malta, was surprised that the Germans she encountered looked nothing like the evil brutes she had been hearing about:

*'The Germans looked - how strange – like other men.'*

By October 1914, most Germans in Australia were interned and subjected to various social, legal, and economic restrictions. German foods like lager and frankfurters were frowned upon, and music by German composers such as Beethoven and Wagner, was banned. German-owned shops and homes were attacked. At the wharves, workers refused to work side by side with Germans and worse still, German migrants were beaten, spat on, dismissed from jobs, and refused services at stores and restaurants. A number of them had to remove their children from school because of the ongoing taunts, and some even went so far as to change their surnames to more Anglo-sounding names. Likewise, three years later in 1917, the British Royal family was criticised for its German name, Saxe-Coburg and Gotha, and the court was said to be *'alien and uninspiring'*. King George V retorted: '*I may be uninspiring but I'll be damned if I'm alien!*' He changed his family name to Windsor, while another branch of the Royal family, the Battenbergs, became the Mountbattens. In Russia, Saint Petersburg was changed to the less German-sounding name Petrograd, and in the first Christmas celebrated during wartime, Christmas trees were banned by the Russian Orthodox Holy Synod as they were considered a German custom.

Whatever their reason for enlistment, the volunteers were eager to take up their arms and leave Australian shores to fight the enemy. However, as late as October, *'the volunteers are still playing at soldiers but have not gone away yet.'* (Letter from Waldemar to George, 2nd of October 1914)

The training of the volunteers took place in animal pavilions in city showgrounds or in open fields called camps, such as Broadmeadows in Victoria, Enoggera in Brisbane, Mitcham Camp in Adelaide, and Blackboy Hill and Osbourne Fighting Range in Western Australia. Training consisted of grenade throwing and bayonet practice. There were not enough uniforms to go around. In some camps, recruits either had military hats or shirts, and most of them still wore their civilian clothes.

Broadmeadows Camp, in north-west Melbourne, was home to many young Australians who were to storm the beaches at Gallipoli. The Lighthouse Brigade soldiers who stormed the Turkish outpost of Beersheba in 1917 also received their training there. This training camp once covered more than 80 hectares, and held in excess of 8,000 recruits. The area had first been used by the army in 1911 when a training campsite was set up in the vacant paddocks during Easter. By 1915, it was clear that the war would not be over by Christmas and the camp took on a more permanent state. Wooden huts were built to replace the tents, uniforms replaced civilian clothes and real rifles started being used by the recruits, instead of the saplings or broomsticks utilised for drilling practice.

Usually the next destination after the Broadmeadows Camp was Egypt for further training of soldiers, before they were sent to the Dardanelles or to other battlefronts later in the war. By an ironic twist of history, the camp was even used as a temporary home for Melbourne's first wave of Turkish immigrants in 1969.

A number of soldiers and officers that distinguished themselves during the war owe their training to Broadmeadows Camp, among them, Albert Jacka, who was awarded Australia's first Victoria Cross of the war for retaking a trench position the Turks had captured. Hailed as the symbol of the Anzac spirit, he also fought in France where he won further honours. After the war, he made a name for himself as a businessman and was also elected Mayor of St. Kilda Council.

The daily routine in camps was very regimented: roll call was at 6:00 am, with parade starting at 6:45 am for 45 minutes. Then, breakfast consisted of black coffee with no sugar or milk, dry bread, and bully beef. Parade resumed again at 9:00 am for three hours, followed by a lunch of dry bread, stew, and black tea. At 2:00 pm, another parade would take place until 4:30 pm, followed by dinner of black tea, jam, and bread. Lights out was at 10:15 pm. Some nights, the volunteers were given free time to go to town, write letters home, watch concerts, or attend boxing tournaments held in camp.

A number of recruits were undisciplined and unruly; they did not address their superiors properly or chose not to stay overnight in the camp but returned home every evening. A number of revolts occurred in camp, mainly against the unappetising food, lack of availability of leave, and the stress of separation from loved ones. Some of the new recruits just marched off as they had expected to be taken to the battlegrounds straight away and not 'waste' their time in camps.

It is no wonder then that the news of leaving for overseas was welcomed with excitement and cheers. This notwithstanding, a handful of men actually deserted while saying farewell to their dear ones, just before boarding their ship.

Charles embarked from Fremantle on the *Ascanius* on the 2nd of November 1914. A day earlier, Turkey had entered the war. He arrived in Egypt on the 3rd of December, where he settled with his unit at Mena Camp at the foot of the majestic pyramids. Before his departure, he had become engaged to his Australian girlfriend, Gertie.

Around the same time, in early November 1914, 38 Australian and New Zealand ships, escorted by *HMAS Sydney*, sailed from Western Australia with 20,000 troops looking out for the German cruiser *Emden*, which had already sunk 23 Allied ships, including the Russian cruiser *Yemtschuk* and the French destroyer *Mousquet*. On the 9th of November 1914, the *Emden* attacked the British radio and telegraph station at Cocos (Keeling) Islands. The *HMAS Sydney* intercepted the *Emden* and, in the battle, the German vessel ran aground on a reef with most of its crew losing their life in the Indian Ocean. The *Emden* captain and officers, including a relative of the Kaiser, Prince Franz Joseph of Hohenzollern, were captured and interned in Malta. Earlier in October, another Australian naval expedition had succeeded in capturing New Guinea and the Bismarck Archipelago from German hands.

These initial success stories helped to boost morale and encourage more men to volunteer. As Waldemar put it in a letter he sent to George in October:

*'The war news seems to be very favourable of late so people are getting calmer.'*

However, from the very beginning, it was clear that this was not going to be the easy-to-win war it was purported to be. It proved hard to predict which way the conflict would go. For instance, in the battle of Marne, between the 5th and 12th of September 1914, the Germans were stopped from advancing. Earlier, in Mons, between the 23rd and 24th of August, the British had held

*35. Waldemar, Christmas 1914*

the Germans but were later forced to retreat, as the Germans occupied the town. Thus, during these early days, a stalemate had already been established, which would be the *'ordre du jour'* for the rest of the war. The network of trenches, known as the Western Front, which, by the end of 1914, stretched over 600 km, from the Channel Coast in Belgium to the mountains in Switzerland, did not do much to break this deadlock.

What had started as the third Balkan war had spilled over borders and thus became a European war. The bloodshed would continue to spread even further, making this a global war. Many wrongly thought the conflict would be over by Christmas. As things turned out, the war ran for over four Christmases and only finished six weeks before a fifth one. It would take a bit longer for the Kaiser's troops to reach Paris, if they would ever do so.

Although there have always been conflicts and wars since the advent of mankind, yet, all of these paled in comparison to what the world was about to witness. At the outbreak of war, the British Foreign Minister, Sir Edward Grey, shocked by the turn of events, coined the often quoted phrase that captured the doomed spirit of the era so well: '*The lamps are going out all over Europe; we shall not see them lit again in our lifetime.*'

The worst was yet to come.

CHAPTER V

# *In the Land of the Pharaohs*

***'Dear Homefolks,***

***Well we are getting used to living, moving and having our being in a sandy sea of sand. We have got our tents now and things are getting a bit more comfortable.'***

*(Letter from Bert Smythe, 3rd Battalion, 18th December 1914)*

Perhaps one of the most iconic images of the First World War is the photograph that captures the members of the 11th Battalion, in all their uniformed glory, sitting on the steps of the Great Pyramid of Cheops in Giza.

For most Anzacs, this was the first time they had set foot outside the shores of Australia. They were awed by the grandeur and majesty of the Pyramids. Thousands posed proudly in their uniforms, either sitting stiffly on camels or together in a *'photo de famille'* with their entire battalion on the steps of these Wonders of the World.

Being the cultured and educated gentleman that he was, Charles would have been excited to be in one of the ancient cradles of civilisation, and to be so close to his family in Malta. His family received a photo showing him sitting proudly on a camel, with the majestic Sphinx as a backdrop. This was his last known photo before he sailed to meet his destiny in Gallipoli.

The City of Perth Regiment, or as formally known, the 11th Battalion, which Charles joined, was among the first infantry units raised for the A.I.F during the First World War, and the first battalion recruited in Western Australia. Together with the 9th, 10th and 12th Battalions, it formed the 3rd Brigade.

*36. Charles in Egypt*

*37. Charles as Private 157, 11th Battalion*

Between 1914 and 1918, this battalion took part in various battles: Gallipoli in the Dardanelles, and on the Western Front at Somme, Bullecourt, Ypres, Menin Road, Polygon Wood, Broodseinde, Poelcappelle, Passchendaele, Lys, Hazebrouck, Amiens and Albert.

The regimental colour patch is a rectangle made up of two bars, the top one brown and the bottom, light blue. The badge is a seven-pointed star, surmounted by an imperial crown. In the centre, within a band bearing the name of the regiment, sits a voided outline of a swan on top of a ribbon bearing the motto '*vigilans*' (watchful), which text is positioned above the Roman numerals XI.

Raymond Lionel Leane was the Captain and Company Commander of the battalion. His four brothers also fought in the war and between them, they raised nine sons who fought in the First or Second World War. After Gallipoli, Leane went on to lead the 48th Battalion in France.

One of the most colourful characters of the battalion was Private 951 David John Simcock. Before enlisting, he owned a successful fruit shop in Fremantle, which he later moved to Perth, calling it '*The bloke with the pink top*', a reference to his distinctive red hair.

It was on the 10th of January that the 11th Battalion commanding officer Lieutenant Colonel Lyon-Johnston suggested a group photo on the Great Pyramid at Cheops. He called out for Pink Top, and when the man was found posing in the lower part of the pyramid, he was ordered to go to the very top, where he could be seen standing to attention.

Charles was also well liked by his mates. He was admired for his excellent language skills: his English was impeccable and he also spoke French, Italian, and Arabic. Because of his linguistic abilities, he was appointed as an interpreter.

On the 3rd December 1914, Simcock, Charles, and their battalion had arrived in Egypt after a five week voyage from Western Australia. The narrow passage separating Princess Royal Harbour and King George Sound at Albany, from where ships crowded with troops sailed for Egypt and the Dardanelles, is now called Atatürk Entrance, after the Turkish commander in Gallipoli.

Any enthusiasm they might have felt when at last their ship sailed towards Europe quickly evaporated as boredom on the long, tedious journey set in, not to mention the seasickness. There were complaints about the canteen facilities, including the restrictions on beer, as well as the distribution of mail. Mostly, they killed time writing letters and gambling, playing the 'crown and anchor', 'two up' and cards. Simcock took the opportunity to practise his entrepreneurial skills on board by coming up with various money-making schemes such as reselling goods from the ship's canteen at a profit! He also had a generous side as with the money he made, he bought a number of oranges from various native boats surrounding the vessels when it docked at Aden, which he gave to his fellow soldiers free of charge.

Two and a half years later, in a letter to Mr. E.J. Hayes, Waldemar gave a very detailed picture of life on a troopship, the rough seas, the crowded quarters, and the endless tedium:

*'Just a few lines to let you know that I am well. I am writing this letter under very trying conditions. The sea is getting very rough, so much so that writing letters is becoming very arduous work. The ship is tossing and heaving about like an infuriated animal. Of course, this is to be expected as we will be in the Bay of Biscay in a day or so, and it is mid-winter over here.'*

*'On account of the ubiquitous submarine, lights are forbidden to be shown from any of the ships, so that you can imagine what life is like. Fancy about 500 men herded together in a space of about 25 by 30 yards, all the portholes tightly closed and the men lying about on the floor or in hammocks tightly packed like sardines. The acrid smell of humanity that strikes one on going below is an odour that will not be easily forgotten. However, it is marvellous how soon one becomes used to all sorts of smells, especially if one's olfactory organs are not abnormally developed.'*

*'It is very monotonous on a troopship. The same thing from day to day, and not even a bit of land or a stray ship to break the monotony. Of course, there are other ships besides ours. In fact, it is a fairly big convoy. Unfortunately, that production of modern war, the censor, is not kind of heart, so I am unable to mention how many or what sort.'*

In fact, for security reasons, Waldemar was also unable to divulge his route and destination.

*'We spend most of the time on board about the decks reading books. Now and again the tedium of such a life is enlivened by a concert or a boxing tournament. In the shape of work we get an hour's physical drill per day. This is known as 'physical jerks.' Then there is the unavoidable guard, and such like things. However, this latter event does not happen more than once a week, so I do not think one has much to growl about in the shape of work.'*

The military ship *Ascanius*, carrying the 11th Battalion, had first anchored at Port Said, and then sailed on to Alexandria. Following this, the troops had to endure a five hour train journey to Cairo, and after that, a further one hour tram trip to Mena, near the Great Pyramids, where the Anzacs had set camp. This was only a few minutes' walk away from Cheops, the oldest and largest pyramid, and the Sphinx. The headquarters of the Australian Division were housed at the Mena House Hotel. The bustling city of Cairo was only about 16 km away and could be easily reached by train.

It would take some time for Egypt to grow on the troops after their initial disappointment that their first destination was not the war arena where all the action was happening, but to a hot and arid place for yet more training! Some were scared that the war would be over by the time they reached Europe. In letters sent to loved ones, the soldiers complained about the conditions: *'Sand everywhere in your tucker, in your eyes'* and they called Egypt *'a sandy piece of land'*.

Slowly but surely, the troops started to get a taste for the adventure and novelties that Cairo had to offer in abundance: the colours and smells of the bustling markets, and even the allure of exotic ladies adorned with sheer veils and blinding jewellery.

The local populace saw the arrival of the Australian troops as a god-send; soldiers meant a booming business. The friendly outgoing Aussie character soon won over the Egyptians and everyone, including the youngest children, came up with various schemes to make money. The produce on sale varied from water to sweet juicy oranges. And of course then there was the flesh trade. The area Haret el Wasser became known as the Wazzer red light district, which was dotted with black crosses on white signs denoting out of bounds areas. Egypt soon started to be known as *'the land of sin, sand, shit and syphilis'*. A number of soldiers suffering from venereal diseases were repatriated to Australia. As Corporal Hedley Kitchin of the 6th Battalion wrote in his diary: *'If there is any town in the British Empire or Europe which beats Cairo for immorality, then it must be bad indeed.'* Kitchin would take part in and disappear on the day of the Gallipoli landing.

The Australian and New Zealand troops started to gain a bit of a bad reputation. After they outgrew the novelty and excitement of the pyramids, museums, palaces and the zoo, they soon became restless and bored. They invented all kinds of pranks to pass the time, such as hiring donkeys from the natives to race each other, knocking baskets of oranges off the hawkers' heads, or even driving a tram at full speed after getting rid of the poor driver. Some of them used to stay out all night, drinking, gambling, and tasting the pleasures of the 'joy houses', sometimes even without paying, and never returning to camp. Sometimes, parades had to be cancelled as not enough men turned up in the particular battalion. A few soldiers took time out during training to smoke a pipe or chat to a nurse.

If, at first, the Egyptians had welcomed the troops with open arms, they had become tired of being the focus of their practical jokes or the victims of racist behaviour. One must keep in mind that these Australian young men were the by-products of the White Australia policy introduced after Federation, which was still very much in vigour at this time.

In the same letter to Hayes quoted above, Waldemar referred to this unruly behaviour of the troops who also acted up whenever they went ashore on their way to the battlefields:

*'By the way, I forgot to tell you that wherever the Australians land, the place is put under martial law, and all the pubs are consequently closed. That was the reason why no leave was granted before 12. The pubs close at noon when the Australians are about, so there you are. Still, with all the precautions taken, men were and did get _____!'*

Little respect was shown to the officers, especially if they were arrogant or pompous. Some men refused to salute them, made up rude songs about them, and assaulted military police. Australians were of the opinion that as long as they did their duty on the battlefield, what they did off the job was their concern. They also did not believe that the parade ground drill improved a man's fighting prowess. Their argument was that they were enduring very tough combat drills and lengthy route marches in the sand. They worked hard, but they also played hard. Now that we know what lay on the horizon for them, we find it easy to forgive them for these transgressions. After all, they were only letting their hair down and enjoying their time to the full before they would be called to do their duty.

After January 1915, more discipline was imposed on the troops. Conditions also improved after February and, as a result, their behaviour changed for the better: they were allowed a day off every week and only trained for eight hours every day, on weekdays. However, as late as October 1915, some of the Anzacs' comportment in Egypt still left much to be desired. In a letter from General John Maxwell to the Australian Governor General, while on one hand, he praised the bravery of the Australian fighters, he conceded that in Egypt *'they get impatient, and the officers have no real hold on their men. The men do not see the necessity for either training or restraint.'*

The Australians reciprocated the prejudice of the English towards them. They considered the English to be arrogant cowards in comparison with the colonials. In their eyes, the Brits were not much good at fighting. In the subsequent months, the English army would be accused of disorganisation, and lack of initiative and spirit. An Australian officer affirmed '*I wish Kitchener would send us soldiers not boys to do the business.*'

Trouble loomed again on the night of the 2nd of April 1915, which was Good Friday. Big fights erupted in the Wazzer district after several New Zealanders, who had been infected by prostitutes with venereal disease, decided to get their revenge. Entering a brothel, they went on a rampage, breaking everything, burning furniture, and throwing bedding out on the street. They cut the local fire brigade's hoses and assaulted the firemen. The British military police opened fire, killing one soldier and wounding one Maori and four Australians. In turn, the New Zealand soldiers verbally abused the Red Caps, and attacked them with rocks, beer bottles, and lengths of fire hose. This riot, which would be followed by a similar outbreak in July 1915, was described as *'the greatest bit of fun since we have been in Egypt.'*

The men continued to grow restless and bored in Egypt; they just wanted to go and fight. Their prayers were quickly answered. The day after the riots, the long-awaited news finally arrived: departure from Cairo was imminent and some units were given orders to pack for the Front. Leave was stopped and on Easter Sunday, troops started to leave Cairo by train, on to Alexandria.

Charles and his regiment missed all this 'excitement' as, on the 2nd of March, they had sailed on the *Suffolk*; destination, the Greek island of Lemnos. Their departure happened amidst the roar of bands and the cheers of their mates who were less than happy to be left behind. They reached the island three days later, but had to stay on the ship due to shortage of both water and accommodation. The vessel became the troops' living quarters for the next seven weeks.

Their days at Lemnos, which is located about 80 km away from Gallipoli, were spent on long marches and climbing hills to get fit. As they got closer to the fighting, everyone was in high spirits and brimming with excitement, despite remaining unsure about the destination.

For a proficient writer of letters, Charles's correspondence at this time became scarcer and scarcer. He sent the photo of him on the camel to his father, who in turn wrote him two letters dated the 9th of April. The last two letters from Charles were dated the 8th and 21st of April, respectively. In the first one, he complained of not getting letters from home for two months, although he had received six from Gertie, while he wrote the second one at sea on the way to Gallipoli. This letter contained only one line saying he was in good health.

The 3rd Brigade, of which Charles's battalion formed part, was chosen to be the covering force for the landing in Gallipoli with the task of seizing and holding the high ground to the north-east of the landing area. The moment they had long been waiting for had arrived at last! This was going to lead them to their moment of glory. Boys were going to become men. The adventure had begun!

# CHAPTER VI

# *The Dardanelles Debacle*

***'Expeditions which are decided upon and organised with insufficient care generally end disastrously.'***
*(British Prime Minister David Lloyd George)*

In the book, *'Greek Fire'*, a biography of the operatic diva Maria Callas, author Nicholas Gage reports that, in July 1959, Sir Winston Churchill was a guest on the infamous Onassis's yacht *Christina*. When the yacht was cruising close to the Dardanelles, Churchill expressed the wish to disembark at Gallipoli to pay tribute to the fallen soldiers. However, his wife, Lady Clementine, insisted that this might not be such a good idea. She felt that the visit would be too emotional for the 84 year old, and the yacht passed by the Peninsula at night while her husband was sleeping.

No wonder, for Churchill had been the first Lord of the Admiralty in the First World War and as such had played an important role in the disastrous Gallipoli campaign. This was one of the hardest-fought but futile episodes of the war, and remains an indelible black mark on the pages of Britain's otherwise almost impeccable military history.

On the other hand, for Australians, Gallipoli came to signify a new page in Australia's history. Military-wise, it was a minor campaign, but for Australians, it is considered a major event in their history, a sort of coming of age ritual that marked the beginning of the emergence of this young nation as a country in its own right, only fourteen years since its federation. The campaign also helped to change the opinions of the highest echelons of the British army on the character of the Aussie soldier: from an undisciplined, anti-authority figure to a loyal, courageous hero. The tradition of the Anzacs was well and truly established and immortalised once and for all on this sacred battleground.

Before the war, very few Australians would have been able to locate Turkey on a world map, and fewer still would have heard about Gallipoli. This peninsula, with the Aegean Sea to the west and the Dardanelles Straits to the east, is an important waterway that had played a strategic role throughout the tumultuous history of the region. It was here that the mighty Persian king Xerxes launched his invasion of Greece. The epic Trojan War brought to life in Homer's Iliad was also fought at the entrance to the Dardanelles.

Turkey entered the war in October 1914, when it aided two German warships, the *Goeben* and the *Breslau*, to bombard Russian Black Sea ports. The Allied Forces rallied together to minimise the Turkish threat on Egypt and protect Russia. Control of the Dardanelles would give Britain and France a direct sea route from the Mediterranean to the Black Sea, and from there they would be within easy reach of their ally Russia. The Black Sea, vital for

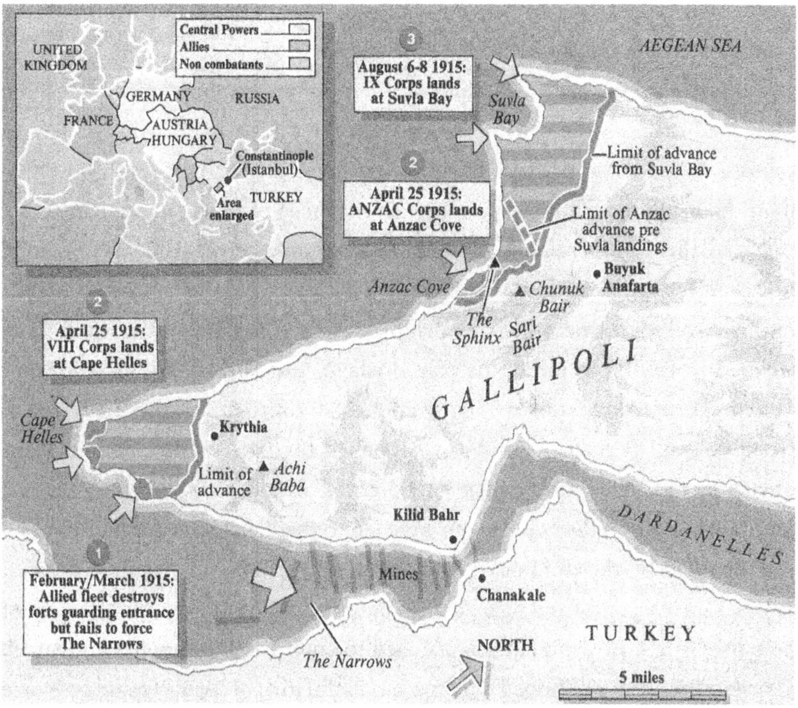


*38. Map of Gallipoli*

Russian trade as it provided the only access to the world's trading routes, was a highly-contested area, that had also been at the centre of the Crimean War between 1853 and 1856. In that conflict, the British and the French had united with Turkey against Russia, mainly to curb the latter's ambitions in taking over Ottoman territories and in its dispute with France over holy places in Palestine. However, old loyalties are frequently forgotten and more than half a century later, Britain and France came to Russia's aid against Turkey.

The Allies' plan was to conquer the Dardanelles, which would open the way to Constantinople, about 300 km inland. This enigmatic and exotic city that evoked sultans' palaces, harems, and eunuchs had eluded the cream of European nobility more than once, and had fallen into the hands of the Ottomans in 1453. Predictions abounded that marching into Constantinople was just a matter of weeks away, with some bold optimistic souls even venturing to say that it was a matter of mere days! Europe also had eyes on the Turkish oil riches of Mesopotamia or present-day Iraq. The ultimate aim was to finally bring the mighty Ottoman Empire to its knees.

Centuries earlier, in 1645, the Ottoman Sultan Ibrahim, nicknamed the Mad, had ordered the Turkish armada to set sail to Malta with a view to capture it. This was perhaps to vindicate the crushing defeat 40,000 Ottomans had suffered eighty years earlier in the Siege of Malta, at the hands of a much smaller number of knights and locals. Upon receiving the Sultan's order, however, his chief admiral, the Kapudan Pasha, accidently placed a candle on his naval map where the wax drippings fell on the little island until they covered it. Failing to locate the prized island, the Turkish admiral declared *'Malta yok'* or *'Malta does not exist'*, and sailed off to attack the Venetians in Crete instead! Perhaps, if this had happened in 1915 to the Gallipoli maps owned by the Allied forces, then *'Gallipoli yok'* would have prevented the loss of thousands upon thousands who were led to their slaughter for nothing.

Frustrated by the deadlock in the West, the British War Cabinet decided to open up a second front against Germany's new ally, Turkey. On the 13th of January 1915, a British plan was hatched to enter the Dardanelles and attack Constantinople. It was hoped that this successful turn of events would convince neighbouring neutral countries, such as Bulgaria and Greece, to join up. This joint effort would push the Austrians out of the war, thus leaving Germany to fend on its own.

The attacks in the Dardanelles started on the 19th of February 1915, when four Turkish forts at the entrance to the Straits were destroyed by British and French warships. A month later, on the 18th of March, the Turks strengthened their position and attacked, sinking three battleships. As time went by, and the naval attacks did not yield many results, mainly due to bad weather, the old warships, and the heavily-mined Straits, it became more and more evident that military backing for the naval operation was paramount. In order to make any leeway, the Navy needed the help of the troops who would land on the beaches of Gallipoli and capture the forts guarding the entrance. The original plan of just sticking to a naval battle was scrapped. Lieutenant General Sir William Birdwood, commander of the Australian and New Zealand forces, pushed the idea of attacking the Peninsula on land.

On the 11th of March, the Secretary of State for War, Lord Kitchener, appointed General Sir Ian Hamilton as Commander of the Dardanelles forces. Seven days later, General Hamilton arrived at Lemnos, armed with two travel guides about Turkey and a text book about the Turkish army.

The landing strategy seemed easy on paper: Australian, New Zealand, British, and French troops were to land in seven different places on the Peninsula, on both sides of the Dardanelles, to confuse the enemy. From the start, however, even the least experienced militarist could see this was an ill-planned, ill-equipped, and futile campaign. Gallipoli was a notorious impregnable coast and Hamilton did not even have up to date maps of the area! To be fair to Churchill, he wanted to stick to his original idea to tackle Gallipoli only through a naval operation. Even after the landing, Churchill kept insisting on more direct Navy involvement.

From March 1915, the troops that had been training in Egypt started sailing to Lemnos to wait for their next instructions. In Malta, British ships remained docked on alert in the Grand Harbour, while troops were entertained with a performance of Faust at the Royal Opera House, pending their departure.

Saturday, the 25th of April, was designated as D-day for the landings. The evening before, a fleet of ships sailed from Lemnos port past the island of Imbros, the seat of the Headquarters of the Allied Navy, anchoring for the night.

The troops were ordered to rest as much as they could. They killed the endless waiting time by talking of home and their families, sharing photos, smoking, attending church services and praying. Mixed feelings of excitement, apprehension, and anticipation hung in the air.

A.R. Perry of the 10th Battalion provided a vivid eye witness account of the atmosphere and general mood just before landing:

*'Come on, lads, have a good, hot supper - there's business doing.' So spoke No. 10 Platoon Sergeant of the 10th Australian Battalion to his men, lying about in all sorts of odd corners aboard the battleship Prince of Wales, in the first hour of the morning of April 25th, 1915. The ship, or her company, had provided a hot stew of bully beef, and the lads set to and took what proved, alas to many, their last real meal together. They laugh and joke as though picnicking.'*

The troops were given a big tot of navy rum for Dutch courage. They had conflicting and mixed feelings. On one hand, they felt proud to be finally doing their bit for King and Country, and considered this the ultimate test for Australians to prove to other nations that their young country could fight like everyone else. The world would be judging Australia by their achievements the next day. Others felt that they would not be up to it. Nervously, they cleaned their rifles and sharpened their bayonets.

Towards dusk, leading ships carrying the 3rd Brigade, which included Charles with his battalion, started the journey east. During the night, a British convoy of battleships, destroyers, and transports followed with soldiers from France and other parts of the Empire. The ships arrived off the coast at about 2:30 am. It was so dark, no one could pick out the land. The sea was smooth. The moon had set and the stars shone brightly. Although the air was chilly, the weather was fine. It promised to be a clear day.

Perry continued his account:

*'Then a voice: 'Fall in!' comes ringing down the ladder way from the deck above. The boys swing on their heavy equipment, grasp their rifles, silently make their way on deck, and stand in grim black masses. All lights are out, and only harsh, low commands break the silence. 'This way No. 9 Platoon, No. 10 Platoon C Company.' Almost blindly we grope our way to the ladder leading to the huge barge below, which is already half full of silent, grim men, who seem to realise that at last, after eight months of hard, solid training in Australia, Egypt and Lemnos Island, they are now to be called upon to carry out the object of it all.'*

The troops were transferred onto barges, thirty to forty men in each one, that were towed to steam pinnacles. As the shore came nearer, all the lights went out and an eerie silence prevailed. There was no sign of life or any sound from the land. Like hunters in a forest waiting for their prey, thousands of pairs of eyes skimmed the land for the slightest movement but there was nothing. As the dawn started to break, the sinister cliffs came into view.

The boats were cast off once they approached the shore. At exactly 4:29 am, a small steamboat let off a flame and sparks from its funnel. This alerted the Turks who started firing. The boats were rowed furiously under a barrage of fire from an invisible enemy. In typical Aussie humour, some of the men played cards under fire:

*'They want to cut that shooting out'*, one protested, *'somebody might get killed.'*

*'Then crack-crack! ping-ping! zip-zip! Trenches full of rifles upon the shore and surrounding hills open on us, and machine-guns, hidden in gullies or redoubts, increase the murderous hail. Oars are splintered, boats are perforated. A sharp moan, a low gurgling cry, tells of a comrade hit. Boats ground in four or five feet of water owing to the human weight contained in them. We scramble out, struggle to the shore, and, rushing across the beach, take cover under a low sand-bank'.*

The 9th, 10th, and 11th Battalions formed the first wave to land on the Peninsula. With bayonets fixed, they had to wade through over 25 metres of water high up to their chests, carrying their weapons and ammunition above their heads so as not to get them wet. As soon as they were immersed in the icy water, they went under and then came up, struggling and spluttering water, and charged forward. But there was nowhere to hide. Difficult landscape and rough terrains rose up to welcome them: shrubs, cliffs, tangled valleys, high ridges, forty-five degree slopes, and thick undergrowth.

The plan was that the troops would land close to an open plain, but they landed too far north from the intended landing point where the Turkish army took advantage of the treacherous ridges. The designated site was Gaba Tepe, a little promontory jutting into the sea; however, they landed in Ari Burnu, which was later renamed Anzac Cove.

The exposed beach and steep sandy hills offered no cover to the Anzacs. For some, the heavy weight of their backpacks, with over 40 kg of equipment and magazines, pulled them down below the water and caused them to drown.

As the day dawned and it was light enough to see, the Turkish guns on Gaba Tepe, and from other batteries, hailed down incessant shrapnel on the landing troops. The battleships and cruisers protected the boats carrying the landing parties as best they could:

*'Se-ee-e-e . . . bang . . . swish! The front firing line was now being baptised by its first shrapnel. Zir-zir . . . zip-zip! Machine-guns, situated on each front, flank and centre, opened on our frontline. Thousands of bullets began to fly round and over us, sometimes barely missing. Now and then one heard a low gurgling moan, and, turning, one saw near at hand some chum, who only a few seconds before had been laughing and joking, now lying gasping, with his life blood soaking down into the red clay and sand.'*

*"Oh-h!' comes from directly behind me, and, looking around, I see poor little Lieutenant B——, of C Company, has been badly wounded. From both hips to his ankles blood is oozing through pants and puttees, and he painfully drags himself to the rear. With every pull he moans cruelly. I raise him to his feet, and at a very slow pace start to help him to shelter. But, alas! I have only got him about fifty yards from the firing line when again, bang-swish! and we were both peppered by shrapnel and shell. My rifle-butt was broken off to the trigger-guard, and I received a smashing blow that laid my cheek on my shoulder. The last I remembered was poor Lieutenant B—— groaning again as we both sank to the ground.'*

Upon landing, the soldiers hurriedly unburdened themselves of their laden packs. The battalions became intermingled in the melee and confusion that ensued. Many soldiers lost sight of their leading officer and charged aimlessly forward, while being pounded by an unseen and relentless foe hiding in the thick shrubbery. Father J. Fahey, a Catholic priest assigned to accompany the 11th Battalion at the landing, together with other chaplains, comforted the dying and wounded, without taking notice of denomination.

Colonel Walter Cass of the 2nd Brigade reported that *'practically all semblance of company and battalion formation was lost.'* He also testified to the bravery shown:

*'At least 90 per cent of them had never been under fire before and certainly 95 per cent had not been under shell fire. Yet they looked as if they were used to this sort of thing all their lives.'*

Colonel Cass went on to take charge of the 7th Battalion, and played a pivotal role in the evacuation of the Peninsula. He died in 1931 following an operation to his appendix.

Strategically, the landing was a complete disaster. However, although the troops had arrived in an unexpected and strange place, they were not fazed. They regrouped and reorganised themselves and pushed forward as much as they could, putting into practice what they had learnt in training: to always keep advancing and to find an officer to follow.

The confusion that reigned supreme on the day of the landing would characterise the rest of the Gallipoli campaign. The lack of direction was not only experienced by the Allied side. Even the Turks were confused and as the campaign wore on, they also started to question the decisions of their military leaders. An unnamed Turkish officer quoted in Robert Rhodes James' book *'Gallipoli'* laments:

*'The last commander of this regiment has been killed. Most of the officers are dead or wounded. I do not even know the name of the hill I am on. I can see nothing. I know nothing. I implore you for the safety of the nation to appoint someone who knows the land and us.'*

Perry went on to describe how everyone helped each other the best they could by lifting the heavy, water-soaked packs. Then, the order came to push forward:

*"Hurry up, there,' says our sergeant. 'Fix bayonets.' Click! And the bayonets are fixed. 'For-ward!' And away we scramble up the hills in our front. Up, up we go, stumbling in holes and ruts. With a ringing cheer we charge the steep hill, pulling*

*ourselves up by roots and branches of trees, at times digging our bayonets into the ground, and pushing ourselves up to a foothold...'*

Lance Corporal George Mitchell of the 10th Battalion aptly described the unfolding scene: *'The key was being turned in the lock of hell.'* As they climbed over the hill, Turkish snipers hidden in the shrubbery took their aim. Those who survived this onslaught went over the hill and down to what was later called Shrapnel Gully and up the other hillside, where they found that members of the 3rd Brigade had started to dig in.

Perry slowly moved down to the beach where the Australian Army Service Corps was landing stores and water under constant shrapnel. The empty barges were then utilised to carry the wounded to the hospital ships. Perry was put aboard the troopship *Seang Choon*, which had the 14th Australian Battalion aboard. The latter were supposed to disembark the following morning. However, they were ordered to land straight away, due to the high number of casualties.

*'And so, after twelve hours' hard fighting, I was aboard a troopship again — wounded. But I would not have missed it for all the money in the world.'*

Private Keith Wadsworth, whose diary was published in *'The Herald Sun'* of the 20th April 2005, admired *'the blue water sparkling in the morning sun'*, which reminded him of Portsea...*'and I felt I could sit on that beach forever, if only the wretched business would stop.'* He wrote about *'horrible sights'*:

*'Here a poor chap shot in the stomach, there a boy with a face as white as marble and a little puncture in his chest, here an Indian dismally contemplating a bandaged stump where his hand had been blown off. I saw some of our chaps fire at a movement in the bushes lining the edge of the dry creek bed, and one of our men came out writhing and screaming 'Don't fire, you b...... b..., don't fire!' and fell over into the ditch.'*

This was one of many instances of wounding or killing by friendly fire as the thick vegetation rendered visibility difficult.

Private Wadsworth was injured in the shoulders, and numbered among thousands of wounded transported to Malta: *'Malta was a very nice place, and I had a good time there till my wound healed up.'* He had formed part of the same battalion as Charles, the 11th. After arriving at Mudros Bay in Lemnos in the beginning of March 1915, they had stayed there until the 24th April, sailing for Gallipoli on the *HMS London*.

As part of the 3rd Brigade, the 11th Battalion played a pivotal role on the 25th April; being the covering force for the landing, it was the first ashore at around 4:30 am. On the left flank, they were carried north of Ari Burnu Point, and with water up to their waist and dodging bullets, they had to cross around 200 metres of sea. Walter C. Belford, in his book *'Legs Eleven'*, gives a detailed description of the battalion's first taste of fighting on the Peninsula:

*'The 11th Battalion boats all made for the extreme left, to the north of Ari Burnu. The Battalion had to get past the point of Ari Burnu, where the beach receded and consequently had about 200 yards more to row than the other boat parties. They thus came under heavy fire while still on the water and many casualties were inflicted before the troops reached the shelter of land.'*

*'Dark(ness) and confusion'* reigned. Those that managed to dodge the bullets and land were *'sodden with seawater and their equipment dragged'*. The so-called 'shelter of land' provided no protection at all:

*'...a machine gun was enfilading them and causing many casualties, and the unfortunate wounded and killed rolled down to the beach below'*

*'...the 11th Battalion suffered heavily in the first day's action.'*

The battalion continued to be at the forefront throughout the Gallipoli campaign. Ten days after the landing, one of its companies mounted the A.I.F's first raid of the war against Turkish positions at Gaba Tepe. Subsequently, the battalion was heavily involved in defending the frontline of the Anzac beachhead. In August, it made preparatory attacks at the southern end of the Anzac position before the battle of Lone Pine. The 11th Battalion continued to serve on the Peninsula until the evacuation in December.

After the withdrawal from Gallipoli, the 11th Battalion returned to Egypt. It was split to help form the 51st Battalion, and then was brought up to strength with reinforcements. In March 1916, it sailed for France and the Western Front, where, until 1918, it took part in bloody trench warfare. The battalion's first major action in France occurred at Pozières in the Somme valley in July. After Pozières, the battalion manned trenches near Ypres in Flanders before returning to the Somme valley for winter.

The 11th Battalion helped to stop the German spring offensive in March and April 1918, and later that year participated in the great Allied offensive launched east of Amiens on the 8th of August 1918. This advance by British and Empire troops brought about the greatest success in a single day on the Western Front, one that German General Erich Ludendorff described as '*the black day of the German Army in this war*'.

Operations of the battalion continued until late September 1918. In November 1918, after the Armistice was signed, members of the A.I.F began to return to Australia. In February 1919, the 11th and 12th Battalions were amalgamated due to steadily declining numbers in both battalions. They remained so linked until their last members returned home for demobilisation and discharge.

Charles's name is listed in the appendix of Belford's book as a member of the 11th. However, although he was present at the landing, in the following months, much to the chagrin of his family and his beloved Gertie, no one seemed to have a clue as to his whereabouts. He had just disappeared. His sister wrote to him on the 29th of April, singing Gertie's praise. The family had received a letter from her, announcing that she and Charles had got engaged before his departure. Everyone loved her picture and congratulated Charles on his choice. Charles's sister also wrote about the torpedoing of the French cruiser *Leon Gambetta* by an Austrian submarine on the night between the 26th and 27th April 1915, and mentioned friends she knew who were among those who perished. Other letters from his family were returned to them, stamped with the dreaded words '*Missing*'.

Besides Charles, there were other Maltese who played their part in the Gallipoli campaign, even though they did not have their own battalion. Some 1,000 Maltese volunteers of the Maltese Labour Corps and a number of Egyptians, along with some older recruits of the British Army Service Corps, were brought in to relieve the soldiers from work at the beaches and piers. They undertook stevedoring duties at Gallipoli, offering fresh water to about 24,000

Allied soldiers from barges that sailed from Malta to Anzac Cove. Supplies, such as biscuits and tinned meat, the staple food for the troops, were also provided by the labour corps. Even as late as September 1918, they would also serve in Salonika (now Thessaloniki), where the French and British had landed on the 25th of September 1915 to help the Serbs.

The Maltese Labour Corps was in addition responsible for constructing and maintaining roads with the engineers and tunnellers, as well as building bridges and dugouts. The eminent war historian, Charles Bean, recorded that they suffered greatly in the conditions of Gallipoli and that foolishly, their camp had been set up in a position on North Beach, where it was open to enemy sniping and shelling.

When I visited Gallipoli in November 2000, I stumbled upon the grave of Giuseppe Camilleri, a twenty-seven year old member of the Maltese Labour Corps. He is buried in Ari Burnu cemetery, which lies between the beach and the cliff under Plugge's Plateau, about one kilometre north-west of the Lone Pine Cemetery and Memorial.

It must have been a bittersweet moment for those Maltese fighting or helping out in Gallipoli. This was history giving a big slap in the face to Turkey. Just like the Turks had attacked Malta in the siege of 1565, it was now Turkey's turn to be invaded. In their editorials, the dailies in Malta aimed at the Maltese sense of pride and patriotism by stressing this almost karmic twist:

*'In the ages to come the story will be told of the capture of the Dardanelles and the fall of Constantinople, with the consequent expulsion of the Turk from the soil of Europe which he has too long defiled. Thus Malta will have shared in the work of finally destroying its age-long foe.'* (*'The Daily Malta Chronicle'*, 26th of March 1915)

***39. Grave of Giuseppe Camilleri, Maltese Labour Corps***

Other Maltese worked in Salonika as builders for the British military. Some of them succumbed to malaria and are still buried there. Since they did not die directly because of the war, their surviving families were not entitled to a pension. Second Lieutenant Arthur Henry Vella, who had transferred from the King's Own Malta Regiment of Militia (KOMRM) to the 6th Royal Irish Fusiliers, is buried in Salonika, after being killed in action in 1917 when he was only twenty-two years old.

Maltese were also found within the Royal Navy. As '*The Daily Malta Chronicle*' reported:

'*There is indeed hardly a British ship of war that does not comprise some Maltese among its gallant crew.*'

Furthermore, a few Maltese were taken as prisoners of war by the Turks. In his 1922 book '*Il-Kbira Gwerra Dinjija*' ('*The Great War*'), Emilio Lombardi records the fate of three brothers from Senglea: Ġwann, a priest, Ġużeppi, and Duminku Xerri. While Fra Ġwann worked as a nurse at the Holy Sepulchre Convent in Jerusalem, his two siblings were working in a Turkish factory in another part of Palestine. When Turkey entered the war, they were arrested and taken as prisoners of war in a Turkish camp. They were treated very badly and forced to work like slaves, with little food and water. Duminku who was twenty-six years old, got weaker and weaker and one day, did not have the strength to work. He was refused medical help, despite his brothers' pleas for mercy. Instead, the guards butchered Duminku with their bayonets in front of his brothers.

A distinguished Maltese lawyer, Lewis F. Mizzi, also ended up interned by the Turks. He had played an important role as negotiator of various treaties for the Turkish sultanate. His work earned him various decorations, accolades and awards, mainly from Persia, Spain, Great Britain, Romania, Greece, and Turkey. Moreover, he was the consular representative of various countries, including Spain, Greece, and Romania, as well as the proprietor and editor of two English newspapers: '*Levant Herald*' and '*Eastern Express*'. His property was confiscated and, upon his return to Malta in 1920, he donated a collection of minerals for display at the Museum of Natural History in Mdina.

The conflict in the Dardanelles had indeed brought the war close to Malta. From February 1915, as a backup navy base, the harbour in Valletta was kept busy with the daily arrival of British and French warships, on their way to Gallipoli. A number of drydocks workers sailed with them to work on the ships, as well as labourers forming the Maltese Labour Corps. Doctors and nurses from Malta had also been transferred to the Western Front as early as August 1914. Besides serving as the main imperial navy dockyard and supply base in the Mediterranean, Malta was also an important mail transfer station during the war.

The French troops always made a good impression with their elegant and colourful uniforms. French officers used to be the special guests of Maltese families for dinners and musical evenings at home. Their memorable arrival and presence in Valletta were recorded in Captain Albert Laferla's diary entry for the 5th of March, quoted in John Mizzi's book:

*'Strada Reale never presented such an animated scene as today. The French sailor with his characteristic red tassel in the centre of his cap and the French soldier were in evidence everywhere. The officers wore gorgeous uniforms of striking colours. No drab khaki.'*

Captain Laferla in his diary entry of the 12th April also wrote about the arrival of 500 Maoris to train at the beachside location of Għajn Tuffieħa. He was impressed with their strength and imposing physique. He also described the haunting sentimental music of a traditional love song that filled the night air:

*'...the Maoris around the tent rested their foreheads in the palm of their hand and thought of those they had left behind. Many will never see them again.'*

But perhaps Malta's biggest contribution to the war effort was its role in welcoming thousands of wounded and convalescing soldiers from May 1915, for which it was endearingly dubbed as *'the Nurse of the Mediterranean'*. It was formally declared as a base hospital in March 1915 as Egypt's hospitals had not been coping well with the unexpected large number of casualties. While Cyprus was considered too close to the conflict, Gibraltar was deemed too far. The choice thus fell on Malta, which reprised the role it had played during the Crimean War sixty years earlier. After the evacuation from Gallipoli, the numbers started to decrease. However, later, Malta served as a hospital for the casualties from Salonika.

In one of her letters to Charles, which would never reach him, one of his sisters mentioned that: *'Today lots of wounded are expected in Malta from the Dardanelles.*' From Sydney, Lady Edeline Strickland, wife of Malta-born Governor Strickland, supported the Red Cross and took special interest in the care of the thousands of Australians convalescing in Malta.

The first 1,200 wounded from the Dardanelles arrived on the 4th of May 1915. From June to July 1915, Malta received an average of one hundred wounded Anzacs every single day. In the beginning, there was only limited accommodation, only four hospitals with a total of 268 beds. But, as the number of wounded grew dramatically, many public and private buildings, such as schools and even convents, were converted to hospitals. By September, Malta had 13,700 hospital beds accommodated in various makeshift hospitals. Convalescent depots sprouted all around, from Cottonera and Ricasoli in the far south, to the northern tip of Mellieħa, and Fort Chambray on the island of Gozo.

Queen's Hall in Old College Street in Sliema, which was previously a music hall, was converted into a home for soldiers suffering from shell shock. Where previously soft melodious music had filled the building, now all one could hear were the screams and groans of people in agony, and the wards teemed with patients staring into space, oblivious to their surroundings. The rooms had also undergone conversion into cubicles with padding for protection.

Australia Hall was built to provide entertainment and refreshments to those convalescing in the Saint Andrew's, Pembroke, and Saint George's areas. Although sadly the building is currently abandoned, the Australian coat of arms still stands at the main entrance, with the following dedication: '*This building was erected in November 1915 by the Australian Branch of the British Red Cross Society for the benefit of the Soldiers of the Empire.*'

The Maltese, whose hospitality is even documented in the Acts of the Apostles, as well as being linked with the Hospitallier Order of the Knights of St. John, competed with each other as who would be the first to greet the wounded, hug them, and offer them tea, coffee, and other amenities. They mobilised and organised themselves to make the stay of the wounded in Malta as pleasant as possible.

The medical course at the university was suspended so students could volunteer their services. Many women, who had never worked outside their home, performed nursing duties, while schoolchildren helped out by rolling bandages. Boy scouts read newspapers to soldiers and ran errands, such as posting letters or carrying X-rays or specimens to laboratories. Those who owned cars took the recuperating men for drives around Malta, showing them the historical sights, and lectures on Maltese history and customs were delivered to the troops. A tea room was opened at the Malta Union Club premises in Għar id-Dud, Sliema run by the Ladies Committee of Saint John Ambulance Association and the British Red Cross to offer *'harmless refreshment and hospitality to the wounded'*.

Later on, when these men started to feel better, they would to be invited in groups of eight or ten to Maltese homes. When those that recovered had to go back to the battlefields, their Maltese friends saw them off back to their ships. Daily examples of kindness were reported by the Maltese newspapers, and donations of cash and in kind kept coming in. Even British expatriates did their bit: a Mrs. Lawson offered to entertain a group of wounded every Tuesday to tea in her house, while a Mrs. Arthur Snagge offered the use of her carriage two or three times a week.

Friendships were struck between the Anzacs and their Maltese hosts. The fact the local populace spoke English eased the settling in of the convalescing troops. This notwithstanding, a few isolated violent episodes did occur between the convalescing men and the locals. This even led the Governor to establish a disciplinary compound at Għajn Tuffieħa to deal with these 'black sheep'.

Throughout the war, the Maltese press published countless letters and articles full of appreciation and testimonials from Anzacs about the kindness of the Maltese:

*'Hospitality may well be adopted by the Maltese as their national motto. Past history and present generosity give them right to that claim.'* (A.J. Vernon, 13th May 1915)

One Australian soldier even sent his war diary to *'The Daily Malta Chronicle'*, which, when published in June 1915, became the first Gallipoli diary to be published anywhere. On the 28th May 1915, the same newspaper had published an eye witness account from a wounded soldier convalescing in Malta who spoke with pride about the gallantry and heroism shown at the landing. He spoke of boats almost losing all their occupants and returning to the ship to get more. The general mood was one of optimism and the perception was that once the troops had landed, the rest would be easy.

The grandfather of Anne Quinane, the former Australian High Commissioner to Malta, was one of the thousands of Anzacs who enjoyed Maltese hospitality and care, after he was wounded in his left eye at Gallipoli. Private Patrick Joseph Quinane of the 7th Australian Light Horse regiment arrived on the 29th of July 1915 and was transported to St. Andrew's Hospital in Pembroke. He stayed in Malta for about a week and was then transferred to England. Unfortunately, his eye could not be saved and he was discharged as medically unfit in September 1916.

Among the wounded arriving in Malta was Joseph Scotto, known as Joe Scott, of the 13th Battalion, a Maltese who had migrated to Melbourne in 1913, as a beneficiary of the Papaffy emigration fund. It was hoped that the familiar sights of his motherland would give him a speedy recovery. Another convalescent soldier was the Commander of the NSW Light Horse Brigade, Colonel John Arnott, whose father owned the famous Australian biscuit company.

Yet, despite all efforts to heal the wounded warriors, military funerals became a common sight and around 300 Anzacs have their resting place in various military cemeteries around the island.

Another moving and poignant firsthand account of the landing was given to the newspaper by Charles George Needham of the 3rd Field Company Australian Engineers. Needham had a visit from Charles's sisters, Marie and Melita, at the Valletta Hospital. The pair were probably eager to meet someone who might know of their brother's whereabouts for they had not heard anything from him since his last letter dated 21st April on his way to the Dardanelles.

Weeks and months went by, with still no news of Charles. His father got more and more preoccupied. On 18th June 1915, he wrote to Charles saying that in Malta he had met many wounded who knew him and *'they were without exception very bright boys and all spoke highly of you.'* His father also mentioned that *'no more bells (are) to be heard in the town and near hospitals'* for the comfort of the wounded soldiers. Charles's sister Lily was also to take part in a theatre production staged especially for the convalescing troops.

Desperate for news, Mr. Bonavia sent letters to the Army Headquarters in Gaba Tepe, London, and Egypt, and also wrote to Senator Pearce, the Minister of Defence in Melbourne. In Australia, Gertie was also getting desperate and in a letter sent to the Bonavia family in August, she stated she somehow felt that Charles was alive:

*'We are so deeply conscious of each other that we know and feel each other's presence no matter how great the distance between us.'*

She also clung to the hope that Charles could have been taken prisoner. In another letter to the Bonavia family, shortly after the evacuation from the Peninsula, Gertie wrote of meeting a soldier at a convalescent hospital in Albany, Western Australia, who was sure Charles had been wounded in the leg and then taken to Malta, from where he seemed to have disappeared. However, according to this informant, the Secret Service knew of Charles's whereabouts. Several more months of agonised waiting would have to pass for Gertie to know whether her intuition was right or not.

News was also very hazy regarding the fate of Charles's battalion mate, David John Simcock, also known as Pink Top. The daughter of one of his employees at the fruit shop in Perth noticed she did not receive a gift from Pink Top for her birthday. She asked her father why the presents had stopped coming and he disclosed the reason for this: poor Simcock had been killed at Gallipoli. The date of his death is disputed; some accounts mention he died on the day of the landing when he took leadership of the company after the officer-in-charge was shot down. While the renowned war historian Bean agrees with the date, he describes quite different circumstances surrounding his death, saying that Simcock was killed while carrying a wounded man to cover during a Turkish attack on Baby 700. Others, including the Defence Department records, put down his date of death as early May. His shop in Barrack Street, Perth remained open for business until 1979. Whatever the date and circumstances of his death, for years afterwards, he continued to be fondly remembered by fellow businessmen in the fruit market industry and by his numerous customers.

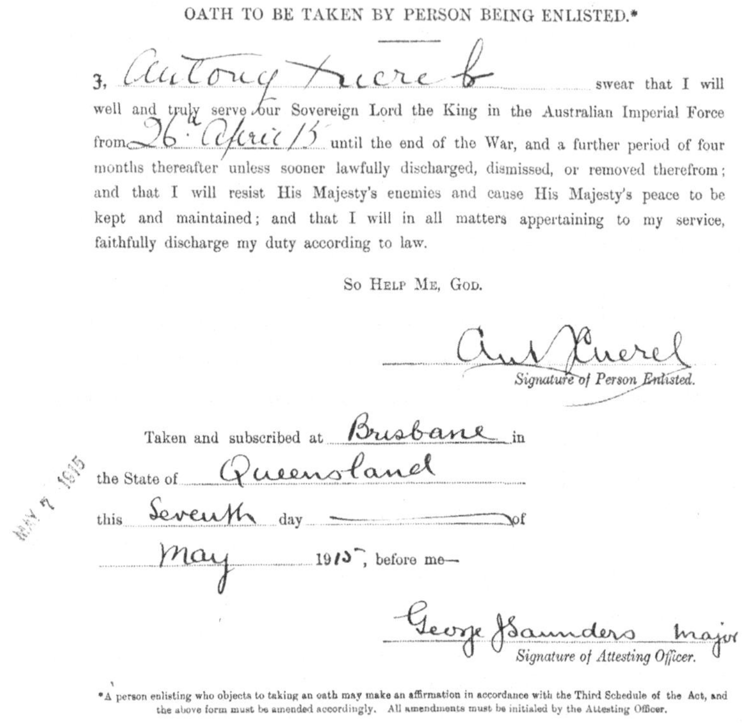

OATH TO BE TAKEN BY PERSON BEING ENLISTED.*

I, Antony Xuereb swear that I will well and truly serve our Sovereign Lord the King in the Australian Imperial Force from 26th April 15 until the end of the War, and a further period of four months thereafter unless sooner lawfully discharged, dismissed, or removed therefrom; and that I will resist His Majesty's enemies and cause His Majesty's peace to be kept and maintained; and that I will in all matters appertaining to my service, faithfully discharge my duty according to law.

So Help Me, God.

Ant Xuereb
*Signature of Person Enlisted.*

Taken and subscribed at Brisbane in the State of Queensland this Seventh day of May 1915, before me—

MAY 7 1915

George J Saunders Major
*Signature of Attesting Officer.*

*A person enlisting who objects to taking an oath may make an affirmation in accordance with the Third Schedule of the Act, and the above form must be amended accordingly. All amendments must be initialed by the Attesting Officer.

***40. Anthony's enlistment papers***

On the 2nd of May, the Australian papers published the first list of casualties and in the following days, the first list of the dead. On the first day of the landing, the Allied troops had suffered 20,000 casualties. It was already evident that the campaign had started on the wrong foot. The Turks held key positions and the Australian leaders, Birdwood and Bridges, requested permission to withdraw.

By early May, on average, most of the officers and half of each battalion had been decimated and some sections had disappeared completely. This stark reality brought the war close to home. A surge in enlistment took place, especially by those who had lost a close friend or relative, whose death they wanted to avenge.

Albert Jacka was the first Australian to win the Victoria Cross for action at Courtney's Post in Gallipoli. The news of a soldier winning a medal also had a positive effect on the number of young men volunteering.

The new recruits were excited and enthusiastic, still not affected by the war and the battle-weary men around them. In June, there were 12,000 Australian new recruits. However, a year later, only half that number volunteered each month. Few soldiers finished the war with the same enthusiasm they displayed upon starting their service:

*'Towards the end of the War, we were so fed up, we wouldn't even sing 'God save the King' on church parade. Never mind the bloody King, we used to say, he was safe enough, it should have been God save us.'*

Newspaper offices had people lining up awaiting news from Gallipoli. The community got organised and set up fund-raising days to collect money for the wounded, raising £312,000 on one particular day. Boys joined the knitting crusade to make socks for the troops and the Lady Mayoress's League published *'directions for knitting two socks at once'* to speed up the production. Care packages with tobacco, chewing gum, warm socks, and any other item that was thought to bring comfort to the troops, were dispatched to the battlefields.

On the 8th May 1915, the British luxury Cunard liner *Lusitania* was torpedoed by a German submarine off the coast of Ireland. It sank in a mere eight minutes, bringing down with it 1,198 passengers. The ship had also been carrying supplies such as cases of shrapnel and boxes of cartridges for the Royal Arsenal at Woolwich. A warning issued from the Imperial German Embassy, printed in American newspapers, reminded readers of Germany's announcement that it would destroy all ships in the waters of Britain and Ireland. The passengers had also received anonymous warnings not to sail. This incident continued to turn public opinion against the Germans and also precipitated the entry of the United States in the war.

On the 19th of May, the Turks launched a massive attack to breach Quinn's and Courtney's posts and to drive the Anzacs back into the Aegean. The bloodiest fighting happened at Quinn's Post, a small ridge of land near the head of Monash Valley from where one had a view of the entire Anzac position.

This attack proved disastrous to the defending army. Ten thousand Turks were either killed or wounded. The Anzacs fared much better, with 'only' 170 dead and 500 wounded.

On this same day, the Australian army lost one of its most colourful characters, Private John Simpson Kirkpatrick of the 3rd Field Ambulance, better known as *'the Man with the donkey'*. Simpson had adopted a stray donkey, which he called Murphy after the circus in England where, before the war, he used to give children mule rides. He was a familiar sight, carrying the wounded on the donkey through heavy fire, down the steep slopes to the beach dressing stations. One of the other donkeys used for this purpose even wore a headband with a Red Cross. Originally, the headband used to be an armband belonging to Colonel Alfred Sutton of the 3rd Field Ambulance who gave it to the donkey so it would be recognised as part of the Unit.

At just twenty-two years of age, Simpson received a bullet in the heart while carrying two wounded diggers and died instantly. The inscription on his gravestone reads: *'He gave his life so that others may live'*. No one knows what happened to poor Murphy!

By this time, less than a month after the beginning of the land offensive, the ground was covered with dead bodies that had been left unattended for weeks. The men ate, drank, and slept surrounded by the flies and stench emanating from the corpses. As time went by, the men got used to death surrounding them; they walked gingerly over the dead bodies and kept going on their way, unruffled. There was no time to be sentimental. They became fatalists, accepting the probability of dying. Some of them even swam in the Anzac Beach waters under the rain of Turkish shrapnel, as if trying to defy death.

Many bodies got buried deeper and deeper in the earth as trenches were strengthened. An armistice was reached and the 24th of May, which happened to be Empire Day, was designated as the one when both sides could bury their dead in peace. At the same time, some apprehension reigned as regards the impossibility of burying the dead without exposing one's position to the enemy.

This was the first time when the two sides would come face to face as 'friends'. Although fraternisation between opposing troops was frowned upon and discouraged, on this occasion, pleasantries and cigarettes were exchanged.

The two sides were aware of the daunting and unpleasant task at hand. As one Turkish captain who supervised the burials so aptly put it:

*'At this spectacle even the most gentle must feel savage and the most savage must weep. God pity all of us.'*

The Australians only chose men taller than six feet to form the burial party to intimidate the Turks into thinking that all Australian soldiers were big men. Major Aubrey Herbert, a Turkish and Arabic speaker, and a member of parliament representing South Somerset, headed the Anzac side during the truce. His brother, the 5th Earl of Carnarvon, would go on to form part of archaeologist Howard Carter's team when he discovered Tutankhamen's tomb near Luxor in 1922.

When the time set aside for the burials lapsed, it was business as usual as troops from each side returned to their respective trenches, no doubt preparing for the next day's offensive.

Although the dead were given a proper burial and the area was cleared of rotting corpses, the living conditions of the troops were not that greatly improved. The trenches lacked proper drainage. The long, hot days of summer brought with them flies and mosquitoes, which caused the outbreak of diseases, which were further exacerbated by poor hygiene. Venereal diseases, a relic from the times spent in Cairo's red light districts, pneumonia, sunstroke, measles, dysentery, enteric fever, diarrhea, paratyphoid, and flu outbreaks proved a common occurrence due to bad hygiene, heat, fatigue, and the concentration of so many people in the confined quarters of the crowded trenches. Rats and mice became the troops' uninvited guests. Flies and lice were their constant companions. Especially from August onwards, the number of men succumbing to sickness exceeded that of the wounded.

Water shortage was an issue. Weeks went by without the soldiers being able to bathe, shave, or wash their flea-infected uniforms. There was no soap and they could only wash in the sea, risking injury or death from Turkish shells.

The sea had also suffered contamination from the corpses of horses, mules, and donkeys. The toilets were very basic, made up of a biscuit tin with two planks across it, a line of them placed side by side behind the trenches, a rather dangerous spot!

Then, in November 1915, the snow and rain arrived. The weather changed from one extreme to the next almost instantly: one moment it was stifling hot, the next, the cold was numbing, there was nothing in between. For most of the Australian troops, this would have been the first time they experienced snow. Their excitement was short-lived, however, as the freezing conditions brought with them new challenges such as frostbite and hypothermia. Their uniforms proved inadequate for combating such cold weather. The trenches became soaked and everywhere was caked with mud and slush.

If these bad conditions were not enough, the men had to contend with meagre rations and the bland daily diet of bully beef and biscuits. Fit, strong men were starting to look like walking skeletons, battle-weary and old. They longed for home and the simple pleasures they had hitherto taken for granted. So, they looked forward to receiving care packages from home, filled with cakes, puddings and warm knitted socks and jumpers.

The soldiers lived in constant danger of shells raining down, snipers' bullets, machine gun fire, suffocating fumes and shrapnel flying around. If you were unlucky to get hit, the basic medical care provided was not very reassuring. Stretcher bearers worked in a highly dangerous environment, darting from one place to the other to collect the wounded under heavy fire and shelling. The injured were carried to the Regimental Aid Post (RAP) for treatment by the battalion's Medical Officer, whereas the more serious cases were taken to the Advance Dressing Stations. Dr. Charles Mattei, whom we encountered earlier as one of the advocates of Maltese migration to Australia, was stationed in Anzac Cove and his medical work there was mentioned in the Despatches as it contributed to the low percentage of sickness amongst the troops of a particular division.

Most injured men could not wait to go back in action. They idealised death and looked at pain as a welcome prize of duty and proof of their manhood. When they heard news of someone they knew who was wounded or had died, they were filled with more ardour to fight and do their bit to replace them and avenge their deaths with the enemy. They saw it as their sacrosanct duty to fight for the Empire.

Some even hid or minimised their injuries so as not to be taken away and miss being in the thick of the action. This happened more in the early days. Later, when the men became battle-weary and could see no end in sight, the wounded became the subject of envy and were offered *'a fiver or a tenner for your wound'*. They looked forward to receive a *'blighty'*, a popular euphemism for a nonthreatening but incapacitating wound, requiring evacuation to Britain, and far away from the hellish Gallipoli terrain.

A soldier's kit was heavy, weighing around 27 kg and included a rifle with a 43 cm stainless steel bayonet with a hooked quillon. The army supplies given to the troops, though, were meagre and of low quality. They had no steel helmets and only received eight rounds of ammunition a day. The handmade bombs, produced out of jam tins in a makeshift factory set up behind the frontline, were too basic and also highly unpredictable, exploding at whim. Many hands were lost when they exploded before being thrown.

And so, the days became weeks, the weeks became months, and the Gallipoli campaign dragged on with no major achievements. Areas, such as Chunuk Bair, conquered by New Zealand troops, were captured only to fall back into Turkish hands shortly afterwards. Towards the end of June, the troops were hit with the realisation that the fight was going to take longer than expected. They grew bored in the trenches just waiting around for the order to charge. Trying to give the trenches a 'homely' feel, they even named their dugouts with such fancy names as *'Ivy Lodge'*. The sense of adventure wore off and only duty and pride kept them going. The only highlights of the day were receiving letters from home and the arrival of newspapers, especially *'The Bulletin'*. They looked forward to mail day more than payday! The men also worked on producing their own monthly magazine, and a newspaper called *'Peninsula Press'* was published daily.

The feelings experienced by soldiers just before battle ranged from expectation to nervousness, tension, thinking of loved ones far away, and a resignation to their fate. Just before charging, some got the attack of the shivers, but once the whistle blew, their mind would go blank and they would just go for it. They felt the fear and did it anyway. After a battle, everything would be just a blur, as if everything had been a bad dream or a hellish nightmare.

In August, a number of significant battles were fought. On the 7th of August, the 8th and 10th Light Horse Regiments attacked Turkish trenches at the Nek. This ridge was around 45 metres wide at the Anzac line, narrowing to about 27 metres at the Turkish front, with opposing trenches approximately 18 metres apart. Horse brigades had been sent without their horses to Gallipoli as infantry reinforcements in May, following the heavy casualties during and after the landing. During the charge, these regiments were almost wiped out. Sergeant Major Walter Strang, a career soldier with the 8th Light Horse Regiment, was shot in the arm and then evacuated to Malta.

A day earlier, the British troops had tried another landing in Suvla Bay on the west coast, and the Australian 11th Battalion attacked Lone Pine. Although the Australians won their battle, both sides suffered heavy losses: 2,000 Anzacs and 7,000 Turks. The great courage shown by the troops resulted in no less than seven Victoria Crosses being awarded. The attack on Suvla Bay was less successful. The 20,000 British troops who landed were inexperienced and although initially they were met with no opposition, they suffered fierce resistance two days later.

These two attacks were part of a plan devised by Hamilton to conquer Hill 971 (also called Koja Chemen Tepe), the highest peak in the Sari Bair range, which was an important observation and gun battery position for the Turks. The plan was that, once conquered, the Allied troops would then go through the town of Maidos, allowing the Allied fleet to sail to Constantinople without any hindrance and forcing Turkey to surrender.

The Allied attack came from four directions involving the New Zealanders, Australian and British troops, the Gurkhas and some Indian Sikhs positioned at Suvla, Lone Pine, Cape Helles, and Baby 700. However, Hamilton's plan came to nothing as the Turks managed to push the attackers back.

Chunuk Bair meant a disastrous offensive for the New Zealanders. They had managed to take it, but with the failure of the other offensives at Suvla Bay and Baby 700, they lost it as the 10th of August dawned, after three days and nights of ceaseless fighting, during which troops were exposed to snipers' fire and artillery bombardment.

By this time, around the middle of August, the hospitals in Malta and Egypt had reached their full capacity and the wounded were dispatched to England. The 27 hospitals in Malta had assisted over 135,000 sick and wounded.

On the 21st of August, the attack to capture Hill 60 was made by crossing an open wide plain in broad daylight exposed to the enemy's fire hidden away on higher ground. This proved to be a recipe for disaster! Shells burnt the dry scrub and the advancing troops caught on fire and were spared from their misery by their mates who shot them.

Following these futile and bloody offensives, Australian correspondent Keith Murdoch wrote a letter to the Australian Prime Minister Andrew Fisher where he referred to the incompetencies of the British command. The Australian media never blamed the Anzac troops themselves for the failure of the campaign. The letter triggered the cessation of Allied 'piecemeal' attacks in Gallipoli and sowed the ground for a formal recommendation to evacuate the Peninsula. The letter also reached the War Cabinet in London, and led to the dismissal in October of Sir Ian Hamilton whose mantra '*dig, dig, dig*' had become part of the troops' psyche. His replacement, Charles Monro, would inherit the mammoth task of organising the withdrawal from the Peninsula.

The animosity between the British command and the Anzac troops was a two-way street. Resentment abounded because the Anzacs were paid more than their British counterparts. While an Anzac soldier received six shillings a day, the British private had to contend with just one shilling. The British also still held memories of the Anzacs' behaviour in Egypt. On their part, the Australians looked down on British soldiers, whom they referred to as chickens. They were also reluctant to respect English officers. The Anzacs' worst animosity was focused on the British Military Police, the so-called Red Caps, who never lost an opportunity to victimise them.

The English complained that Australians were good fighters in the line but bad soldiers out of it. They played cards on sentry duty, dressed carelessly, left trenches untidy, and did not stand to attention when addressed by superiors. Most officers though turned a blind eye as they knew that, in the moment of truth, they would never be let down. The Australians were there to fight; they knew what they had to do and were not prepared to obey useless orders issued only for the sake of enforcing authority.

When the British Prime Minister Asquith had asked whether Australians were good enough to fight at Gallipoli, Lord Kitchener is reported to have replied sarcastically: '*Good enough if a cruise in the Sea of Marmara was all that was contemplated.*'

In 1916, the commander of the British Expeditionary Force, General Sir Douglas Haig, echoed Kitchener's low opinion of the Anzacs due to their unruly and undisciplined behaviour. By 1918, however, the high echelons of the British Army changed their tune drastically: General Haig allegedly claimed the Anzacs were now well disciplined as they always attacked when they were told to. King George V also had words of praise:

'*The Australian and New Zealand troops have indeed proved themselves as worthy sons of the Empire.*'

Other colonials and even some British officers did not subscribe to Kitchener's point of view as they looked up to the Anzacs. On the 8th of May, a British major, watching the advance of the Second Brigade up the bare slope of Krithia Spur at Helles, observed the following about the Anzacs:

'*They were at home in hell-fire... They laughed at it; sang through it. Their pluck was titanic. They were not men, but gods, demons infuriated...Our men tore off their helmets and waved them, and poured cheer after cheer after those wonderful Anzacs.*'

The Gurkhas, the crème de la crème of Indian soldiers, called the Australians '*the White Gurkhas*'. In a letter from Gallipoli, Lieutenant Alex Steele of the 9th battalion, who was later promoted to Major, and was killed in action at Broodseinde in Belgium on 7th of October 1917, expressed his anger at those who had dismissed the Anzacs:

'*Now what do the people in Australia think of the men who were 'only good enough to drink and debauch in Cairo?' I should think that the dirty-minded cowards who ran us down would feel rather small.'*

In September, there was a lull in fighting as that year, Ramadan fell in that month. This served the two sides well as both defender and invader had time to recover from the heavy battles in August. The Turks were showing an unexpected display of resilience and proving to be a formidable foe, coming from a long tradition of fearless and valiant warriors. Their strength and courage also stemmed from their fervent Muslim faith. Perhaps their might and fighting power had been underestimated by the Allies as at the time, the Ottoman Empire was on the decline, regarded as the sick man of Europe. The Turks were led by the German Commander-in-Chief General Liman von Sanders. In February 1919, von Sanders would be arrested in Malta and charged with war crimes, only to be released and repatriated six months later.

The Turkish army was held in great respect by the Anzacs. They admired *'Johnny Turk'* as a hardened and disciplined fighter. Other endearing names used by the Anzacs to refer to their enemy were Jacko, Abdul and Mr. Turk.

Contrary to the testimony of a number of Anzacs, the Turks never tortured or mutilated the body. In the first few weeks of the campaign, the Australians were fuelled by unfounded tales of Turkish atrocities, such as the rumour they would cut their enemies' tongues out; but, after the offensive of the 19th May, when Turkish soldiers attacked the Anzac trenches with no hope of success, inching towards a sure death, the Australians realised their enemies were just as brave and fearless. The Turk gained the reputation of a fair and brave fighter who always followed the rules. They treated the enemy prisoners humanely and '*in the best possible fashion*'. A number of Anzacs had implicit trust in their enemy and even refused to wear gas masks as they believed Jacko would never use gas against them.

As the Gallipoli campaign wore on, the Australians viewed their enemy as family men like themselves. Tasmanian Private J.A. Kidd imagined his victim's wife and children and refrained from shooting a Turk a few metres away. But this kind of attitude was the exception rather than the rule. To others, killing came naturally and they took great pleasure boasting both in their diaries and in their letters home of how many Turks they had bayoneted. Some had the lust to kill and liked to use their bayonets indiscriminately, with bloodlust on their mind. Sergeant H.J. Jackson of the 13th Battalion from Glebe NSW wrote:

*'We charged three hills that night. On the first hill, I bayoneted a Turk who was feigning death, with a few extra thrusts. He was an oldish man and on the first thrust which did not go right home, he tried to get his revolver out at me but failed......'*

This notwithstanding, the Australians knew this war had nothing to do with a personal grudge against the Turks. After all, everyone was fighting to defend themselves. A number of acts of friendships between the two sides were reported during the campaign: Lieutenant T.E. Couzens wrote in his diary that Australians went over to the enemy trenches and exchanged *'bully, jam, cigarettes, etc. The whole business was wonderful and proves how madly unnecessary this part of the War is'.*

Besides throwing food, notes, cigarettes, photos and badges between opposing trenches, friendly greetings of *'Hey Turk'* and *'Hello Australia'* were frequently exchanged between the troops. At one time, the Turks even went so far as to warn the Australians to lay low when a German officer was approaching. Soon after, a machine gun began to shoot from one end to the other. During the ceasefire in May for burial purposes, Australian soldiers came to the conclusion that Turks *'do not seem to be a very bad sort of chap after all'.*

The Australians admired and appreciated this sense of fairness with which the Turks fought. Officers discouraged these friendly gestures between enemy trenches as they did not want their subordinates to forget that this was a real war they were fighting, and therefore, they should behave like real soldiers.

On the 13th of November 1915, Lord Kitchener visited the troops in the trenches. He was given a warm welcome and the men cheered him as he said: *'The King asks me to tell you how splendidly he thinks you have done.'* Besides delivering this personal message of encouragement, Kitchener had another aim behind his visit: he had come to the Peninsula to try to shift opinions. Following smuggled reports by journalists such as Keith Murdoch and Ellis Ashmead-Bartlett that gave the true picture of how the campaign was going, public opinion was turning. Sir Charles Munro, who had replaced Hamilton as Commander in the Dardanelles, was recommending evacuation. The situation was further complicated by the entry of Bulgaria into the war on the side of the Central Powers. This opened a direct land route to Turkey for the Germans to supply heavy siege artillery to decimate the Allied trench network. More Allied troops were also needed on the Western Front to put added pressure on the Germans. Thus, withdrawing troops from Gallipoli to strengthen the other battlefields became imperative.

The prospect of evacuation had first been raised on the 11th of October but Hamilton had resisted the suggestion, fearing damage to British military prestige. He pressed the point that such action was out of the question as it would be considered as a huge loss of face for the Allies; furthermore, he claimed it could not be carried out without serious loss of life. He was dismissed shortly after.

***41. Anthony and his battalion, second left, front row***

Lord Kitchener, whose commanding figure and impressive curled moustache adorned the war's recruitment posters all around the Empire, would drown in June 1916 on his way to Russia when the cruiser struck a mine and sank in a quarter of an hour, with only a dozen or so survivors. His body was never found. This gave rise to a number of conspiracy theories, one of which purported that since the newly-appointed Minister of Munitions (and future Prime Minister) David Lloyd George was supposed to accompany Kitchener on the fatal journey, but cancelled at the last moment, this pointed towards an assassination orchestrated by Britain that wanted to see Kitchener and his allegedly outmoded ideas about modern warfare out of the way. Considering that news of his death was received with great shock in England dispelled this theory quickly.

The impending winter would also have had a bearing on the decision to evacuate. By the end of November, snow covered the whole Peninsula. With soldiers standing in a foot or more of snow for days and weeks without taking off their socks and boots, sometimes their skin, and even toes, came off. Many had to have one or both feet amputated to prevent gangrene from spreading. At times, it was too late even to amputate. Frostbite and trench foot were also common.

When, in early December, the troops heard about the evacuation plans, they welcomed the news. But soon after, the reality sank in that they would be leaving their dead mates behind. During the last days, they lovingly tended to the graves of their friends, placing stones and crosses made from building materials and trench supports. After leaving, a man who had experienced the landing, wrote:

*'It was a sad day for us that the order for the evacuation was issued. Every man of the good old 1st Division has someone whom he honoured and respected, lying in one of those solitary graves at Anzac, the thought of having to leave these sacred spots to the mercy of the enemy made the spirit of the men revolt and cry out in anguish at the thought of it. It has even been said that some of the men broke down and cried...when they heard the order. It drives me almost to despair.'*

The Anzacs remained willing to keep on fighting and even die. They were still determined to win at all costs, adopting the popular catch cry: *'Are we downhearted? No, we will beat these Turks if we have to stay here 35 years.'*

Another common reaction was that their dead mates had paid the ultimate sacrifice for nothing:

*'All that sacrifice, all that labour, all that suffering for nothing at all...the flower of Australia's manhood dead...'*

The Anzacs never blamed themselves for the defeat. They put the blame squarely on the English generals for the whole campaign and, in particular, for the blunder of the Gallipoli landing.

Lieutenant Armitage echoed the thoughts of many when he stated:

*'We all feel extreme disgust at the weakness of Kitchener's army who bungled up the Sari Bair action and fuddled Suvla Bay...inefficient, incapable and badly led troops...if they had been shoved in right, we would have finished Jacko.'*

The brains behind the planning of the evacuation was the Australian Colonel Cyril Brudenell White. 135,000 troops were evacuated between the 10th of December 1915 and the 9th of January 1916. The Anzacs withdrew from Anzac and Suvla on the night of the 18th and 19th December. The rest of the British forces left Cape Helles between the 8th and 9th of January 1916. Not a single Anzac soldier was killed and only five were wounded, two at Anzac Cove and three at Suvla bay.

The evacuation night turned out to be a foggy one, which suited the Allies perfectly as they could slip away under a shroud of mist. While guns from the ships covered the troops still on the Peninsula, surplus stores were destroyed or consumed so as not to leave anything useful for the enemy. Artillery and ammunition, sacks of flour, and food supplies were also tainted, rendering them useless. Sadly, many horses were shot dead or poisoned. Tents were left standing to give the illusion of normality. The remaining troops played cricket up to the last minute to give the false idea that everything was as usual. In typical Australian humour, some food was left behind, accompanied with cheeky notes such as *'have a good feed Johnnie'*. They left the Peninsula saying *'Anybody left?'* and *'Good night Gallipoli!'* All evacuees on the ships met off Imbros Island, 27 km away. The Allies would have to wait until three years later to return to Gallipoli after Turkey's surrender in November 1918.

But perhaps the most genial method used to trick the Turkish army was the self-firing rifle invented by Private William Scurry of the 7th Battalion. A tin was placed at the back of the gun with water dripping in it. The weight of the water in the tin pulled the trigger, giving the illusion that the Anzacs were still there.

By contrast, when, on the 7th of January 1916, the British withdrew from Cape Helles, they suffered significant casualties. One of these was the Maltese Lieutenant Herbert W. Huber of the KOMRM contingent in Cyprus, who was killed on the very morning of the evacuation day from Cape Helles. He was hit through the head and rendered unconscious for fifteen minutes before he died. The Huber family suffered a double tragedy as Herbert's brother, Second Lieutenant Edgar, was also killed in a car accident in Salonika, where he had been serving with the Maltese Labour battalions. Herbert is commemorated on the Helles Memorial, right at the entrance of the Dardanelles.

Like the Anzacs, Huber had also experienced some prejudice from the British military authorities. In his last letter to his father, dated 21st of December, he wrote that he did not hold much hope for a promotion as the British officers would resent being under the command of a Maltese. But he was not prepared to '*hide my nationality as a Maltese to please any of these snobs. I will show them that a Maltese can face the enemy as well as they can, if not better. I shall get a chance someday to make all those that look down upon us to make them hide themselves when they come to be compared to the Maltese soldier, who fights stubbornly for England though not an Englishman.*'

The eight-month long, ill-planned and bloody campaign thus ended. Ironically, the evacuation proved to be the most successful operation of the whole campaign! What had been a disaster from the beginning culminated in one of the most amazing retreats in military history. The commander of the Turkish forces, General Liman von Sanders, in his book '*Five years in Turkey*', published in 1927, which he wrote while captive in Malta, lauded the Allies for '*the withdrawal had been prepared with extraordinary care and carried out with great skill.*'

The evacuation did not mean the end of the war for Turkey, which continued to fight Russian troops in Romania and Anatolia. In the Middle East, it kept up its involvement against the British who, under the pretext of helping out the Arabs in their struggle for independence from the Ottomans, really wanted a finger in the pie of the region's rich oil reserves. Furthermore, in 1915, the Turks, suspicious of the Armenian minority which had a contingent in the Russian army, arrested many members of the Armenian intelligentsia. Between three and four million Armenians were deported and 1.5 million killed.

The statistics from Gallipoli are equally sobering: there were 252,000 Allied casualties, with 50,000 killed. Of these, 8,700 were Australians. Australia had sent 60,000 men to Gallipoli when its population at the time was just under five million. Of the 20,000 that took part in the landing, 2,000 lost their lives on that first day. During a visit to Gallipoli four years after the landing, Charles Bean, the official World War I chronicler, could still see skeletons of men wearing the Australian uniform, bearing the distinctive shoulder patch identifying their battalion. The defending Turkish army lost 86,000 men, with 300,000 casualties.

These might be huge numbers - and they were - but a year later, in France, more were to die in just one week. Gallipoli offered a slight indication of what was still to come on the Western Front, although the powers that be did not seem to have learnt anything to avoid a similar bloodbath. Most of the Gallipoli survivors went on to fight in France and if they managed to survive the war, this proved an amazing feat in and of itself.

Both sides were victims of bad planning and imperialistic ambitions - a high price to pay for a campaign that achieved next to nothing, at a cost beyond all proportion to any territorial gain.

The Gallipoli campaign remains etched in the annals of Australian history and in the hearts of Australians to this day. It is an important milestone in the military heritage of Australia: the courage, strength, loyalty, and resourcefulness of the Anzacs helped Australians forge a national identity. The Australians felt themselves as the proud sons of the Empire, but also as citizens of a nation in their own right.

This constituted the first major military campaign for the young nation since standing on its own two feet after Federation. The landing, in particular, signified a new era in Australian history, as on that day, Australians proved to themselves they were a force to be reckoned with and could hold their own with some of the best troops in the world, independently of their ties with Britain.

Those that had downplayed the capabilities of the Aussie digger based on the impression that they were an unruly, undisciplined, wild bunch, changed their opinion and started to respect the Australian troops. The bravery shown by the Anzacs at the landing made the English officers reassess their judgement of the Australians. Now they could see the courage of these men matched that of any veteran soldier. The Anzacs took their task seriously and fought only for victory. In the words of many English officers:

*'I really believe today will mark an era in Australian history. On this day Australians proved themselves.'*

*'Australia's rag time army is not so bad after all.'*

*'...the Australian soldier is not a soldier, but he is a fighter, a born fighter, each Australian has his separate individuality and his priceless initiative which made him infinitely better than the clockwork soldier. Discipline irks him, he is not used to it, and it is a thing he can never be made to thoroughly understand; every man considers himself the equal of every other man, it is not in his programme to take peremptory orders but he looks upon a request almost as a command.'*

An English expatriate also admitted:

*'Any of Kitchener's men and Territorials look very awkward beggars beside the Australian, besides of duller intellect...the British seem to be all boots and gapes, but the Australian usually looks as if he knew what he was after.'*

On the 25th April 1916, on the first anniversary of the Gallipoli landing, Prime Minister Billy Hughes exclaimed: '*Soldiers! Your deeds have won you a place in the Temple of Immortals.*'

The legend of the Aussie digger was firmly established: on one hand laconic with a wry sense of humour and dismissing authority and pomposity; on the other displaying unflinching loyalty to his mates, family, and country. The war made them intensely British and absolutely Australian. They epitomised the word 'mateship', sticking together through thick and thin, and risking their lives on a daily basis to save the lives of their comrades. Young and inexperienced soldiers fought at close quarters against a tough and formidable enemy. Fiercely protective, their courage was to become the yardstick for generations of diggers in the years to come.

As 1915 drew to a close, and despite predictions, the war still raged on. Heavy fighting took place on the Eastern Front where the Austro-Hungarians fought the Russians, while, on the Western Front, the Germans fought the French. The initial fiery patriotism and dreams were soon replaced by the grim truth: the short campaign, which was supposed to be over by Christmas, was in fact becoming a long-winded, fully fledged, bloody war.

Amidst the intense fighting in the west, a number of small truces between the two sides were reported in December 1915 for Christmas. Earlier, during the May 1915 armistice in Gallipoli to bury the dead, the authorities had shown their disapproval of any kind of fraternisation between enemy troops. Brigadier General G.T. Forrestier-Walker, chief of staff to Sir Horace Smith-Dorrien of II Corps, had declared:

*'Friendly intercourse with the enemy, unofficial armistices and the exchange of tobacco and other comforts, however tempting and occasionally amusing they may be, are absolutely prohibited.'*

The short breaks from fighting in December 1915 were not as successful and widespread as the unofficial truce that had occurred the year before at Christmastime, recorded in Stanley Weintraub's book *'Silent Night: the Story of the World War I Christmas Truce'*.

As the first Christmas of the war approached, several peace initiatives took place, notably the Open Christmas Letter by a group of British suffragists addressed to *'the Women of Germany and Austria'*. Pope Benedict XV had also pleaded for an official truce, asking *'that the guns may fall silent at least upon the night the angels sang.'*

This attempt was officially rejected, but then the troops decided to take matters in their own hands. When, on Christmas Eve, a German message came over the trenches: '*You no fight, we no fight*', the invitation to take a holiday from the war proved too much to resist.

The troops played soccer, sang Christmas carols, exchanged food and souvenirs, and showed each other family photos. An English soldier received a '*pickelhaube*', the distinctive spiked German helmet in exchange for bully beef and jam. On Boxing Day, he heard someone shouting for him from the German side and met the helmet's owner on no man's land. The German appealed to his new English friend to return the helmet to him as he was going to have an inspection the day after. He promised to give it back to him after the inspection. After the truce, during an attack, some soldiers on both sides shot in the air to avoid killing the enemy who, a few moments before, had been their friend.

However, not all the soldiers showed the same enthusiasm for this fraternisation with the opposing side. A gaunt, sallow corporal with a thick, dark moustache and hooded eyes, belonging to the 16th Bavarian Reserve Infantry Regiment, refused point blank to socialise with the enemy: '*Such a thing should not happen in wartime. Have you no German sense of honour left at all?*'

His name was Adolf Hitler.

# CHAPTER VII

# Somewhere in Flanders

***'Do not think that this is a war...... This is not war. It is the ending of the world.'***

*(A wounded Punjabi soldier writing to a relative in India)*

As the world welcomed in 1916, and the war entered its second year, Charles's family and his fiancée Gertie were still grappling with the uncertainty of what could have happened to him. Although they may have held on to the hope that no news was good news, the lack of information about Charles's whereabouts gnawed at their hearts. In the meantime, while Charles's fate remained a mystery, his friends, Waldemar and Anthony, were following their own destinies against the backdrop of the raging war.

Anthony formed part of the wave of troops that had evacuated from the Dardanelles. He had enlisted in Brisbane on the 26th of April 1915, one day after the Gallipoli landing. His attestation papers show that there were a number of issues surrounding the correct spelling of his surname. There are at least two versions: *'Xueres'* and *'Xuerel'*. The latter misspelling could be due to the scrawl of his signature, which made the 'b' of Xuereb look like an 'l'. These mistakes would haunt Anthony even after the war as, when the service medals were issued, the inscription said *'Xuerel'*, a mistake which was rectified through re-issue of said medals:

*'Letter from 1st Military District HQ Brisbane*

*19th December 1922*

*To Base Records, Melbourne*

*Under separate cover, I am returning 1914/15 Star No.6135, British War Medal No. 9889 and Victory Medal No. 9743 for alteration. These medals are inscribed No.1055 Xuerel A, 25th Battalion. Records held in this Office show this man's correct name as XUEREB. Please have the necessary amendment made.*

*Lt. Col.*

*Adm. Staff, 1st District Base'*

*'9th January 1923*

*Headquarters, 1st District Base, Victoria Barracks, Brisbane*

*With reference to your letter 3/66/22 of the 19th December 1922, I have to advise that the inscription on 1914/15 Star No.6135, British War Medal No.9889 and Victory Medal No.9743 has been amended and the medals returned per separate registered post No.3834 for issue to ex No.1055 Lance Corporal A. Xuereb 25th Battalion.*

*Kindly acknowledge receipt*

*Captain o/ic*

*Base Records*

*Note Star, British and Victory amended and dispatched to 1st District Base 4/12/22'*

Another interesting point concerning Anthony's surname is that of all the thousands of men who volunteered for the A.I.F during World War I, he was the only one whose surname started with the letter X.

There is yet another discrepancy in Anthony's recruitment documents concerning his age at time of enlistment, which is declared to be 26 years 11 months. Now, Anthony had been born in September 1888, which, in April 1915, would have made him 27 years 7 months. The age requirement for recruitment was 19 to 38 at the time so there was no reason why Anthony would declare himself to be younger than he actually was. When news of the heroics of the Anzacs had reached the shores of Australia, the number of volunteers queuing at the recruitment offices in the major cities grew longer and longer. It could very well be that the men accepting attestation papers could not keep up with the soaring numbers and due to overwork, it was not uncommon to put down the wrong age or misspell a name.

In his attestation papers, Anthony declared his calling to be that of a surveyor, employed by the Commonwealth Government of Australia. His next of kin was his father, residing at 70 Strada Levante in Valletta. When asked whether he had ever served in the military, Anthony declared he had been a Cable Guard for eight months. Initially, a *'No'* was typed as a reply to this question but then the answer was corrected and initialled *'AX'* by Anthony.

The Cable Guard had been formed in 1912 to protect the telegraph station in Darwin from possible German interference. Its members were all volunteers and although they all came from different backgrounds, they formed a tight knit group. They held regular practice sessions, which included staging an enemy invasion and taking part in rifle-shooting competitions. One time, their duties also included shooting a rogue crocodile that had been terrifying the locals. Their much-loved Captain Lewis was again their leader when they formed the very first contingent from the Northern Territory. His death at the hands of a German sniper in 1916 would devastate them all.

The attesting officer was Major George J. Saunders who also administered the oath to Anthony on the 7th of May 1915. On the same day, Anthony underwent the compulsory medical examination. Upon enlistment, he was described as being 5 feet 4.5 inches tall (164 cm), weighing 115 lbs (around 52 kg), and having a chest measurement of 32-34.5 inches (86 cm). His complexion was dark, with brown eyes and dark hair. His religion was declared to be Roman Catholic. He had two distinctive marks on both his arms, which were *'Theloid'* and vaccination marks. His vision was 6/6 and 6/6. Anthony declared that he was willing to take inoculation against small pox and enteric fever.

*42. Anthony forming part of the Cable Guard, first left, front row*

On 1st June 1915, Anthony was approved and appointed to the 25th Battalion, 7th Infantry Brigade, B Company, as Private 1055. His term of service was determined to be until the duration of the war and for twenty months thereafter. He embarked at Brisbane on the *Asneas* that same month, on the 29th. After two days at sea, Anthony and his battalion stopped in Sydney, where they took up lodgings in animal stalls at the Sydney showgrounds. No leave was granted; yet, all the men jumped the fence and still managed to explore what the city had to offer. On the 7th of July, they left Sydney, arriving in Egypt via Aden a few weeks later on the 29th, where they camped near Heliopolis. On the 4th September, the battalion embarked from Alexandria to go to Gallipoli, stopping for a few days in Lemnos.

The main role of the 25th Battalion on the Peninsula was to relieve the soldiers involved in the heavy fighting in August. On the 2nd November, still in Gallipoli, Anthony was transferred temporarily for survey duty. On the 9th January 1916, after the evacuation, he disembarked at Alexandria from Mudros, and on the 14th March, he arrived at Marseilles to join the British Expeditionary Force. At first, in France, Anthony and his battalion were billeted with local families, living in barns and marching everyday to their training camp. In April, the battalion was transferred to Armentières, where they were involved in rebuilding trenches, which was no easy task due to the damp weather.

Anthony fulfilled his duties as a private from the 7th June 1915 to the 18th September 1916. He was promoted to Lance Corporal on the 19th September 1916 and continued in this rank until the 25th November 1918. Thus, he was one of the many soldiers who fought on both the Eastern and Western Fronts.

If nothing had been achieved in the Dardanelles, the situation was not much different on the Western Front. A familiar, albeit unwelcome sight, greeted the thousands of soldiers who, having survived Gallipoli, were transferred to the battlefields in the west: trenches. Having failed to advance, and in a desperate bid to end the stalemate, the armies began to dig trenches to protect their positions. By the end of 1914, the trenches went from the Channel coast in Belgium in the north all the way in the south to the mountains in Switzerland, a distance of over 600 kilometres. For the duration of the war, the number of casualties was exorbitant as both sides tried in vain to push the enemy back. If Von Schlieffen had planned a war that would move rapidly, what resulted was a war of stalemate.

The trench system was basically divided into three: the frontline, then the support trenches, and further back the reserve trenches, with a series of communication trenches linking the three lines. Communication between HQ and the frontline occurred via pigeons and telephones, while the Germans also used messenger dogs.

Blind alleys were dug to confuse the enemy, in the event that enemy soldiers found their way in the trenches. The maze-like passageways were also confusing for the trench's own occupants, and each trench had a name on a sign post so that soldiers would not lose their way. The zigzag style of construction also prevented the enemy from firing straight down the line.

A soldier who made it across to the enemy's trenches would have to go through coils of menacing barbed wire put strategically in front of sandbags. In the space between the German and Allied lines, the infamous no man's land, one found shallow dug outs aptly called *'forlorn hope'*, which were occupied by night in the hope of gaining advantage the next day. The desolate scenery here was riddled with potholes, barbed wire, mud, skeletons of burnt trees, wounded soldiers, and decaying corpses. In a bid to forget about the dreariness outside, soldiers decorated the inside with photos of loved ones and postcards, which besides bringing a bit of colour, also made them feel closer to home.

In the frontline, the trenches were usually two metres deep and almost as wide. Wooden duck-boards were mounted to keep feet dry from the slushy mud, and a fire-step allowed one to position oneself to fire at the enemy. While the normal distance between the two sides was usually between 450 to 220 metres, in some places the distance was merely six metres!

Each soldier carried an entrenching tool to dig a 'scrape', a temporary protective trench if caught in enemy fire, and to repair a trench damaged by enemy bombardment. Soldiers also had their rifle, bayonet, and hand grenades. The most effective weapon in the trenches was the artillery gun, first mostly used by the Germans, although soon the Allies began to utilise it with great efficacy. Varying in size between small pieces to huge howitzers, it could reach up to 13 km and could fire up to 600 bullets a minute. One of the main reasons behind the stalemate was in fact the use of the machine gun. This being a huge killer machine against enemy machine gunners, the troops remained trapped in their trenches.

The war remained at a standstill. The need was felt for a mobile vehicle which could move with ease. The most significant development in the weaponry used in the war was undoubtedly the advent of the tank, which was used for the first time in September 1916 in the battle of Flers, during the Somme offensive. The tank originally consisted of a cumbersome piece of metal that could hardly move through the rubble and pot-holed ground of the battle sites. The crews manning the tank from inside had to contend with horrible conditions: the loud engine noise and exhaust fumes. They got sick with the rough movement or even lost consciousness. Its first proper and effective use happened during the attack on Cambrai in November 1917.

A rotating roster was set in place whereby an infantry battalion would usually spend a fortnight rotating between the three tiers of trenches and the billets behind. A soldier could spend anywhere between a day and two weeks in the front trench, then he was moved into the reserve lines, and after that, to the rear to rest, have a bath, and receive freshly laundered clothes, before returning to the front. A lot of the time, though, due to the short supply of troops, it was not unusual for them to spend a month or more at the front before being replaced. The most frightening moment had to be *'going over the top'*, climbing up a ladder out of the trench into no man's land at the signal of an officer blowing the whistle, to charge on the opposing trenches. A creeping barrage, a heavy and insistent artillery bombardment, would be followed by the infantry advancing towards enemy lines. Shells were classified by weight or diameter. High explosive shells exploded on impact, while the anti-personnel shrapnel shells exploded in flight and were designed to kill or maim.

The wounded received their first treatment at regimental aid posts in the trenches where the injuries were assessed. Then they were moved to a casualty clearing station behind the frontline where they received proper medical attention and basic surgery. Farm-houses, ruined factories and bombed out churches, that dotted the French landscape, were converted to these stations. If required, the injured were then transported to base hospitals.

Daily life in the trenches was boring. The routine in the quiet areas would consist of sentry duty, trench maintenance, and weapon cleaning. Soldiers tried to catch up on sleep or used the free time to write letters home.

There was no set time for meals. Not that mealtimes were something to look forward to either. Soldiers had to live on a diet of mainly hard biscuits and canned food: everything was tinned from plum and apple jam to stew, pork, beans, and even butter. Christmas was much anticipated as they could look forward to gifts of pudding, courtesy of Princess Mary, the daughter of King George. In the midst of these inhumane conditions, the soldiers' sense of humour still shone through as illustrated in this poem entitled '*Fear*' which was popular at the time:

*'A terror hangs over our heads,*
*I scarcely dare to think*
*Of the awful doom that each one dreads*
*From which the bravest shrink.*
*It's not the crashing shrapnel shell,*
*It's not the sniper's shot,*
*It's not the machine-gun burst of Hell,*
*These matter not a jot.*
*It's a far worse thing than that, son,*
*With which we have to grapple.*
*It's if we see another one*
*More tin of Plum and Apple.'*

This was the only way to remain sane in such insane circumstances.

Just like in Gallipoli, the conditions here were atrocious and inhumane: besides boredom and bland food, the troops had also to contend with huge rats that infested the trenches, their main source of food the corpses that lay everywhere. Soldiers also suffered from 'trench foot', an infection caused by a fungus when one stood for an extended time in wet conditions. Hygiene was not a priority. Weeks passed without an opportunity to bathe or at least change into a clean uniform. They also scratched incessantly as their hair was invaded with lice. The trenches had indeed become graves for the living.

A lot of work was done at night; patrols were sent out to observe and raid enemy trenches and to repair their frontline parapets. Nightfall was also the safest time to recover soldiers wounded in no man's land. Dawn and dusk were particularly dangerous as they were the most likely times for an enemy attack.

The troops now also had a new deadly threat to contend with: gas, which, although prohibited by the Hague Convention of 1899, was used by the French for the first time in August 1914. They used grenades with xylyl bromide, commonly known as tear gas. The Germans took this a step further as they studied and developed the use of gas, which they tried out against the Russians in January 1915, but the extreme cold neutralised its effects. Three months later, in April, they used chlorine during the second battle of Ypres. In December, the Germans moved onto more deadly gases, such as the chemical phosgene, which is twelve times deadlier than chlorine. It took a while longer for the British to start using gas. They used it for the first time on the 25th September 1915; however, the lack of strong wind did not carry the gas over to enemy lines.

There were mainly three types of gas used. Lachrymatory, or, as it is more commonly known, tear gas, caused the victim to weep acutely. Mustard gas, used by the Germans in 1916, proved to be the worst as besides causing seared and blistered skin and eyes, when inhaled, it burned the lining of the lungs and caused liver damage. Phosgene and chlorine filled the lungs with water and caused suffocation.

At first the only protection were pieces of cloth, soaked in water or even urine, that were put on the face. Later, as the use of gas became more frequent and deadly, gas masks were developed, including masks for horses, who like humans, suffered atrociously. However, these offered no guaranteed protection, especially in the case of an attack with tear gas, which caused the victim to weep acutely and not tolerate the mask.

Lance Sergeant Elmer Cotton of the Northumberland Fusiliers provided this graphic description of a gas attack on Whit Monday in 1915, where men writhed in agony, dying a slow and painful death:

*'Propped up against the wall was a dozen men, all gassed. Their colours were black, green and blue, tongues hanging out and eyes staring. One or two were dead and the others beyond human aid, some were coughing up green froth from their lungs. As we advanced we passed many more gassed men lying in the ditches and gutterways.'*

1.2 million soldiers on both sides were gassed, of whom 91,198 died terrible deaths. Even those who survived a gas attack could expect a lifetime of ill health, frequent stays in hospital due to bronchitis, pneumonia, pleurisy, influenza, tuberculosis, cancer, emphysema and, in a lot of cases, premature death.

Far from the war easing, 1916 was the year that saw the most battles. In Verdun, in the fighting between the Germans and the French, 300,000 souls perished and 750,000 were wounded. Up to a million German and British lives were lost in the battle of the Somme. Battles at sea were also heating up. There had been almost no contact between the German and British navies until the Battle of Jutland in May 1916.

Verdun, in northern France, was the longest battle in the war. In the beginning of 1916, General Erich von Falkenhayn decided to attack the fortress at Verdun. Close to the German border, it controlled access to the eastern part of France. The aim was to destroy the French army and thus, bring England to its knees, at the mercy of the Germans, since it would lose its biggest ally and become isolated. Winning Verdun would also mean strengthening the German position on the Western Front. Verdun also meant a lot to the French as it had been the last fortress to fall during the 1871 war between France and Prussia. The place had become the ultimate symbol of French pride.

The battle started at 7:15 am on the 21st of February 1916 and lasted until December, when the French pushed back the German infantry almost to where they had started. Although this was where the Germans used the deadly phosgene gas for the first time, Verdun signified a huge failure for the German army; so much so, the architect of the German plan, General von Falkenhayn, was removed by the Kaiser himself and sent off to Transylvania. His archrival, Paul von Hindenburg, took the reins of the German army.

Even the British high command experienced change after Sir John French was replaced as head of the British Expeditionary Force by Sir Douglas Haig, who remained in charge until the end of the war. On the French side, General Henri-Philippe Pétain took command of the French forces on the 25th February. His cry *'Ils ne passeront pas'*, meaning *'They will not pass'*, raised French morale. However, if Pétain was considered the hero of Verdun in this war, he was much maligned in the Second World War, when he was seen by most of his countrymen and women as a traitor upon his surrender of France to the Germans on the 22nd of June 1940.

Just before the start of the Battle of Verdun, on the 3rd February 1916, the Australian Army wrote to the Bonavia family in Malta, reporting that Charles was *'missing'* and no personal items of his had been recovered. This conclusion was drawn from various testimonies of eyewitnesses recorded in the wounded and missing enquiries conducted by the Red Cross Society.

The statements came mainly from members of the same battalion who had survived the landing. They gave their testimonies from various hospitals in Malta and Egypt where they were recovering from wounds received in Gallipoli. Some of the information given was conflicting, which is understandable when one recalls the confusion and mayhem that had reigned at the landing. Most said they remembered Charles at the landing, but then he just disappeared. Others said they saw him being shot and dying:

*'Bonavia was lying shot through the hip as soon as they landed together on April 25 and nothing more was heard of him. North was lying alongside Bonavia and says he must be dead. Did not see him dead though. Adds that Bonavia is a Maltese.'*

Pte North 931 11 AIF

Victoria Hospital Alexandria

17.10.15

*'Informant knew Bonavia and they were both in the landing on 25.4.15. When*

*they lined up on the beach on the morning of the fifth day afterwards, Bonavia was not there. No one seemed to know what had become of him.'*

Sgt D. Calvin, 20, 11 AIF A Coy

Gezira Hospital Cairo

W.J. Isbister

22.12.15

*'Witness is a clear intelligent man. He last saw Bonavia between 10 and 11 o'clock on Sunday morning 25/4/15. He was quite alright then. He was absent from the reorganisation on 27/4/15 and witness has not heard anything definite of him since.'*

Sgt Retchford

C Coy 11 AIF

O/S Base Giza Cairo

JLK 1/3/16

*'Dark, clean shaven, not tall, he was a Maltese and his family live in Sliema, Malta. He went out to Australia some years ago and worked in Perth and the Lands Department. On arrival in Egypt last year he was given interpreter's job as he could speak Arabic.*

*Informant says he was told by George Dick 11 AIF C Coy at Anzac last October that on the landing on April 25th 1915 at Anzac, he had seen Bonavia killed outright by a bullet. They never went over that bit of ground again, so his body as far as informant knows, has never been recovered.'*

L/Cpl C.D. Simpson 1181

Battalion HQ

Malta St David's 25.4.16

There also seemed to be some confusion as to Charles's rank:

*'Informant says he is wrongly listed as a Lt. and is a private in A Coy, 11 A.I.F'*

Pte J.J. Ryan 258

Hqr. Signal Coy

11 AIF

Malta All Saints 5.10.15

*'Informant positive no officer of that name. There was however an interpreter of that name, a Maltese private. This man was missing early, but informant could supply no details.'*

C.J. Veitch 975

B Coy 11 AIF

Malta Valletta 22.10.15

'*Lt CE Bonavia* (should be private)

*Informant positive no Officer of this name with the Battalion. There was a private, an interpreter, but it is doubtful if he ever landed on the peninsula.*'

P.Clark

11 AIF B.Coy 416

Malta Għajn Tuffieħa 6.11.15

Charles's family took advantage of the presence in Malta of a few of Charles's battalion mates who were convalescing there. His sisters Marie and Melita visited military hospitals with the hope of getting some firsthand news on the fate of their beloved brother. A statement given on the 13th of October 1915 by Private E.J. Warner, who was in Cottonera, Malta, gave the family some hope that Charles might still be alive, although possibly a captive of the enemy:

'*Informant states that he was not a Lt. but private and an interpreter. He has never been seen since the landing. They have heard that two men who were missing then are in Constantinople, so informant thinks there is a possibility of prisoner.*'

Shortly after receipt by the family of the letter advising that Charles was reported as missing, a Miss Nell G. Kinsman, writing from 52 Warwick Square in London, asked the Secretary of the Defence Department about the possibility that Charles was being held as a prisoner of war:

'*Sir,*

*I was notified from Australian Headquarters London last June, that Private Charles E. Bonavia, No. 157, A company, 11th Battalion AIF was reported missing since in action (undated) in Dardanelles.*

*I write to ask you to be kind enough to give me news (should any reach you of him) and if you are aware of any of his company or battalion in action at that time being prisoners of war. That I fear now is our only hope.*

*The smallest detail of his fate will be most gratefully received.*

*By yours truly*

*(Miss) Nell (?) G. Kinsman*'

The letter, dated 12th of February, reached the Department almost two months later on the 3rd of April. One can only imagine how long and heart-wrenching the wait for a reply must have seemed. This would have just added to the misery and despair that Charles's loved ones were going through.

And when the reply, dated 5th of April was received, it was not very encouraging, either:

*'Dear Madam,*

*In reply to your letter dated 12th February concerning No.157 Private C.E Bonavia, 11th Battalion, who was reported by cable message dated 21/6/15 as missing at the Dardanelles on 25th April last, I have to inform you that with the exception of the names of 8 officers and 60 men, no complete list of members of the Australian Imperial Force who were taken prisoner and are at present interned in Turkey is yet available. The Turkish authorities, however, have established a War Information Bureau at Constantinople, and the American Ambassador is advising the Imperial authorities of all obtainable information concerning prisoners etc. Any particulars relating to Australians would be communicated to relatives.*

*The 'Missing' list is revised daily at Alexandria, and it is hoped the Department will shortly be in a position to afford more definite information.*

*I might add that when a soldier is reported missing and cannot satisfactorily be accounted for, a Court of Inquiry is held at a later date to collect all evidence of the case and record an opinion as to whether it is reasonable to suppose he is dead.*

*Yours faithfully,*

*Captain Officer i/c Base Records'*

Then, finally, the news everyone was dreading came in a letter dated 20th of June, sent from the administrative HQ A.I.F London:

*'Dear Sir,*

*It is my most painful duty to have to inform you of the receipt of advice that No.157 Private Charles Emanuel Bonavia, 11th Battalion, AIF has now been reported 'Killed in Action' at Gallipoli on the 25th April 1915.*

*Any of his effects that may be received will be forwarded to the next of kin who will ultimately receive intimation from the Authorities in Australia, should there be any monies due to the estate of the late private Bonavia.*

*Enclosed herewith is an instructional notice in reference to application for pension.*

*Our most fervent sympathy is extended to you in your bereavement.*

*Yours faithfully,*

*Lieutenant*

*Commanding Inquiry Subsection*

*Administrative HQ AIF'*

A Court of Inquiry, held at Flêtre in France on the 10th of April, had pronounced Private Charles Bonavia as *'killed in action'* on the 25th of April 1915 on the Gallipoli Peninsula. Yet, the family received this information more than two months later. His service record does not include any information on the burial place since, sadly, his body would never be found.

In its edition of the 4th of July, *'The Daily Malta Chronicle'* with *'very much regret'* announced that *'Mr Charles Bonavia, son of Mr and Mrs E.T Bonavia, who was reported missing over fourteen months ago, is now officially reported killed in action at Gallipoli on the 25th April 1915.'*

The newspaper went on to detail Charles's migration to Australia, as well as his enlistment and service:

*'The late Mr Bonavia was one of our promising Land Surveyors and Architects. After graduating at the Malta University, he proceeded to Australia where his talent secured for him an important position under the Commonwealth Government. On the outbreak of the war, however Mr Bonavia gave up his billet*

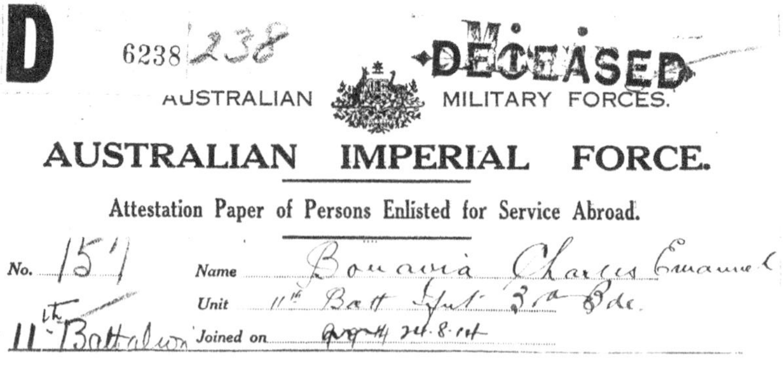
D

6238 238

DECEASED

AUSTRALIAN MILITARY FORCES.

AUSTRALIAN IMPERIAL FORCE.

Attestation Paper of Persons Enlisted for Service Abroad.

No. 151 Name Bonavia Charles Emanuel

Unit 11th Batt Inft 3rd Bde.

11th Battalion Joined on 24.8.14

**43. *'Deceased' stamp on Charles's record***

*and enlisted as a Private. He sought the first opportunity to go to the Front for he joined the first Australian Expeditionary Force which sailed for Europe some months afterwards, and which soon made its name famous for all time. Mr Bonavia was with the splendid Australians on that ever-memorable day of April 25,1915, when the greatest landing in history was effected. It is fourteen months ago, but the dauntless deeds displayed by the sons of Australia, who numbered and still number many of our countrymen amongst them, can never be forgotten.'*

In the article, the newspaper speculated about the date of Charles's death and suggested that he might not have been killed during the landing *'but from the wording of the announcement of his death, it would appear that our gallant countryman fell whilst participating in one of the irresistible attacks which dislodged the Turco-Germans from their all but impregnable positions.'*

Before extending its sympathy to the bereaved family, *'The Daily Malta Chronicle'* proclaimed that *'the late Mr Charles Bonavia has indeed died a hero's death!'*

Various newspapers in Australia also listed Charles as *'killed in action, previously reported missing'*, under such headings as *'Western Australian casualties at the Fighting Front'*. (*'The Daily News'* of the 30th May and *'The Weekly Times'* of the 3rd June 1916)

Thus, the life of yet another promising young man, who still had so much to live for, was cut short. Hundreds of thousands more would join him in numerous other bloody and senseless battles that would do nothing to unlock the stalemate.

One such battle, which also took place in 1916, was that of the Somme in the north of France. This area was chosen for its position on an open and flat plain.

***44. Charles's medals***

If the weather was good, there was no risk of it becoming muddy, as in Flanders. This offensive raged between the 1st of July and the 18th of November, at which time the prevalence of snowstorms and mud brought everything to a halt. The Allies captured an area of circa 125 $km^2$ but did not succeed in breaking through German lines. Their artillery bombardment on enemy positions had not done enough damage, and the infantry marched towards the enemy, who mowed them down with machine guns. The Germans had been in the region since 1914 and used their familiarity with the terrain to their advantage.

On the very first day of the battle, July 1st, the British Army suffered the bloodiest day in its entire history, with 57,470 casualties and a further 2,152 missing. Thousands of others would die on both sides as the two armies fought on for another five rainy months. Most soldiers at the Somme were inexperienced and young and thus, taking part in their first battle.

The losses for the Anzacs were equally tragic. In the battle of Fromelles, on the 19th July, in which the German lieutenant Adolf Hitler took part, 5,500 Anzacs were killed in one single day. Furthermore, between the 23rd of July and the 3rd of September 1916, more Australian blood flowed on the Pozières battleground, with around 10,000 Anzacs dead. This battlefield was later purchased by the Australian government and is now Australian territory. Even on the Western Front, the conflict between the Anzacs and the Allied leadership came to prominence as the Australians blamed the British for the unnecessary carnage.

Anthony and his 25th Battalion saw action in these battles. In the middle of June, they had left Armentières and travelled to the French-Belgian border. On the 29th of July, in Pozières, they were involved in an attack on the German trenches which left a number of them trapped in the barbed wire in no man's land. There was no mercy for them from the German machine guns and bombs that mowed them down. On the 2nd day of the following month, men from the 25th dug trenches for twelve hours, while constantly under enemy fire.

Fast forward to November 1916, when on the 14th, Anthony was wounded in action after he received a gunshot wound in the face at Rouen. Two days later, he was admitted to the 38th Casualty Clearing Station in France and after a few days, he was moved to Rouen. On the 26th November, he was discharged to Étaples, and four days before Christmas, on the same day his friend Waldemar left Perth on his way to England, he proceeded to join his unit, and then his battalion two days later. In a medical report dated 17th June 1920, there is mention that Anthony spent *'one week in hospital in France Rouen with shell shock in 1916.'* Since no date is given for this, it is not clear whether this related to the same casualty of the 14th of November or another one altogether.

News of the huge loss of life in these bloody battles was received with mixed and opposed reactions back home: on one hand, more men started to join up to avenge the death of a mate or relative. On the other, pacifists and radicals increased their fervour in speaking out against the war. In particular, the Socialist movement did not see any benefit coming from the war for the working class. Rather, the conflict would make rich countries richer, with little or no regard for workers.

Lord Kitchener died in a drowning accident in June 1916, but his menacing stance in recruitment posters with the words *'Your King and Country want you'* had started to lose its impact way before his untimely demise. In view of the dwindling numbers of volunteers, the British Prime Minister David Lloyd George introduced conscription in January 1916 for single men aged between 18 to 41 years of age. Later, this also applied to married men. In France, conscription was *'de rigeur'* as well. Pressure was mounting on the Australian and New Zealand governments to follow suit.

In a letter to his sister-in-law Vira in 1916, Waldemar wrote that conscription was not a matter of *'if'* but *'when'*:

*'Nissoponi li dal-waqt jieħdu lil kulħadd għall-gwerra. Hawn ordni li kull ġuvni ma jistax joħroġ mill-Awstralja mingħajr il-permess tal-militar. Ftit xhur ilu ħafna marru l-Amerka sabiex ma ikunux jistgħu jeħduhom il-gwerra. Allura issa ħarġet din l-ordni. Jekk qatt joħduni, jien nista' immur fl-ambulanza għax għandi ċertifikat.'*

*('I suppose that soon everyone will be conscripted. There is an order whereby*

*every young man cannot leave Australia without the military's permission. A few months ago a lot of people went to the United States to avoid being taken to fight. So now there is this order. If I am ever conscripted, I can go with the ambulance since I have a certificate.')*

The conscription campaign divided Australians. Many women joined the Sisterhood of International Peace to campaign against it. The churches were also divided on the issue: while the Protestants were in favour, the Catholic Church was against. The time leading to the referendum kindled fiery debates, street fights, emotional speeches and propaganda.

Writing to his brother George on the 28th of August 1916, Waldemar made it clear on which side of the debate he stood:

*'I think conscription is getting very near. The Cabinet has not come to a decision yet. Most of the Labour members seem death against any sort of compulsion. But I do think that they will turn round to Hughes' point of view. Anyhow the thing has got to come and so I suppose the sooner the better.'*

The anti-compulsionists, led by the Australian labour movement or unionists, argued that white workers sent to fight would be replaced by cheap imported labour. As a perfect example of the ironies of history, John Curtin, in his unionist role, opposed conscription in 1916. However, twenty-six years later, during his stint as Prime Minister at the time of the Second World War, he introduced the very same thing he had previously opposed so vehemently!

Pro-conscriptionists argued that if Britain lost the war, the Kaiser would be the one to determine whether white Australians remained or not. Thus, both poles had the upholding of the White Australia policy close to heart. Maltese immigrants were specifically targeted as they were not considered *'white enough'*. They were refused work and trade unions enforced bans on them. By order of Prime Minister Hughes, the Secretary of State for Colonies was urged to ensure no passports for travel to Australia would be issued by the Maltese Government to single men of military age or to married men not accompanied by their wives.

The arrival in September 1916 of the P&O steamship *Arabia* in Sydney with ninety-seven Maltese men on board was seen as evidence by the unionists of the Government's intentions to replace men gone to war with cheap imported labour. Then of course there was the notorious incident of the *Gange*, detailed in Chapter II, when in October 1916, 214 Maltese labourers and farmhands, some of whom had served side by side with the Anzacs at Gallipoli with the Maltese Labour Corps, arrived in Melbourne. The Government panicked that such an arrival on the eve of the referendum would wreak havoc in regard to its conscription campaign. Some newspapers even went so far as to suggest there was some German interest behind this so-called *'Maltese invasion'*, to blow the chances of conscription.

In June 1916, New Zealand was the first to introduce conscription. Researcher Mark Caruana discovered that Louis Cremona, who had migrated to New Zealand in 1906, had withdrawn his application to enlist in protest against the treatment given to his fellow countrymen on the *Gange*. During his appeal, heard in front of a Military Service Board in May 1917, when the Board argued that he was in fact a British subject, Cremona queried why a number of Maltese had been refused entry into Australia, if they were considered to have the same rights as the British. The Board replied:

*'That has nothing to do with us. If you are aggrieved with the Australians, that is another matter.'*

The newspaper *'Wairarapa Daily Times'* reported that, after the Board advised Cremona to withdraw his appeal, granting him an exemption until the 30th of May, Cremona *'seemed quite pleased, and left the box smiling'*. He seemed very happy with the outcome as he had succeeded in making his point by highlighting the injustice meted out to his countrymen.

The Australian referendum in October 1916 was defeated by 72,500 votes. In December 1917, a second referendum was held. This time, the defeat was even greater: 166,600 voters confirmed a resounding 'No' to conscription. By that time, more and more stories of bloodshed and unimaginable atrocities and butchery on the battlefield had reached the shores of Australia. Telegrams bearing bad news and visits by the local priest or vicar to the families of perished soldiers had become a daily occurrence. The war that was supposed to be over by Christmas 1914 had dragged on for over two years. People were starting to question long and hard the decisions of their government and of the high echelons of the Allied armies. The patriotic feelings of fighting for King and Country had started to wane dramatically.

Meanwhile, the death of the Austrian Emperor Franz Joseph on the 21st of November 1916 brought down the final curtain on the old Europe. It had been the assassination of his nephew and heir, the Archduke Franz Ferdinand, that had triggered the spark that ignited the war. However, if the death of the nephew had started the war, the death of the uncle did not signify its end. Indeed, the hostilities were far from over.

# CHAPTER VIII

# *Waldemar's War*

***'Dear Barbara***

***I hope to be able to meet you tonight at Guildford Station about 7... If I am not there do not wait. I have no time to write. Had an awful time as the prisoners are going away to Rockingham so that ours was the last guard. Wald'***

*(A note dated 19th of December 1916)*

In his numerous letters to his brother George, there is no mention whether Waldemar had heard about the tragic fate that had awaited his friend Charles on the shores of Gallipoli. Although Waldemar had remained in Perth, the war was never far from his mind. He frequently referred to what was going on and gave his take on developments.

In a letter dated 28th of August 1916, Waldemar mentioned to George that he had given a lecture to the local stamp society. Among those who attended were the Honourable Johnston, an inspector of the Western Australian Bank, described as *'an old philatelist'*, and a certain Foster, who *'belongs to the Royal Philatelic Club of Sydney and is the head of an insurance company.'*

*'The Daily News'* of the 19th of August reported on Waldemar's lecture:

*'The programme for the evening: lecture by Mr. Beck on the stamps of Malta was proceeded with the lecturer outlining the history of the island, and also the bringing into operation of the postage stamps of the colony. The lecture proved very interesting to the large number of members present, and at the close of the lecture Mr Beck placed on the table for the perusal of members two books of Maltese stamps, each issue being mounted on separate sheets with a description title to each page explaining how each stamp came into existence with all minor varieties, the whole being beautifully arranged. Messrs. Foster and Johnston*

*speaking to a motion of thanks to Mr. Beck, endorsed the feeling of the meeting when they said that the collection was the best and neatest they had ever seen on any society table.'*

Waldemar sent his brother copies of the write-ups on his presentation:

*'I am enclosing two papers containing the account of the show. It went splendidly and they have already booked me for 2 nights during the ensuing year.'*

However, these other planned lectures would never take place as, by the following year, Waldemar would be in France. He had decided to enlist in September 1916. A letter he wrote to his brother explaining the reasons for joining up was never received as, on the 6th of November, the ship carrying it, the P&O liner *Arabia*, had been torpedoed by a German submarine about 180 km from the southernmost tip of mainland Greece. So we, like Waldemar's family, will never know what pushed Waldemar to answer the call:

*'I am sorry that a letter went down in the Arabia. I explained all my reasons for enlisting. It was a letter I particularly wanted you to get. I do not suppose it is any good going over the same ground now. But still I am really sorry.'* (undated letter)

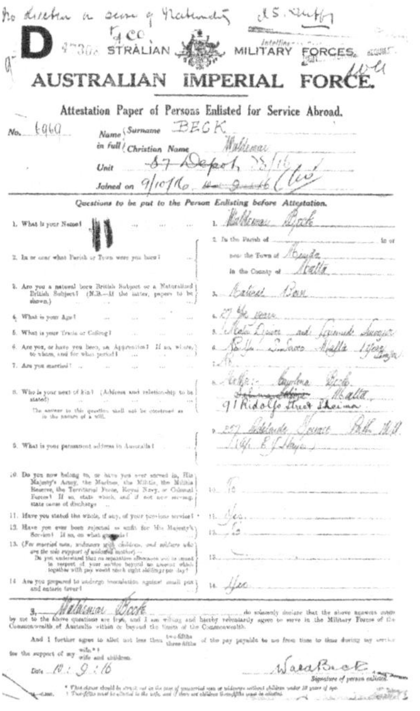
D

STRALIAN MILITARY FORCES.

AUSTRALIAN IMPERIAL FORCE.

Attestation Paper of Persons Enlisted for Service Abroad.

No. 6969 Name in full Surname BECK
Christian Name Waldemar
Unit 87 Depot, 23/16
Joined on 9/10/16

Questions to be put to the Person Enlisting before Attestation.

1. What is your Name? — 1. Waldemar Beck
2. In or near what Parish or Town were you born? — 2. In the Parish of ... in or near the Town of ... in the County of Malta
3. Are you a natural born British Subject or a Naturalized British Subject? (N.B.—If the latter, papers to be shown.) — 3. Natural Born
4. What is your Age? — 4. 27 years
5. What is your Trade or Calling?
6. Are you, or have you been, an Apprentice? If so, where, to whom, and for what period?
7. Are you married? — 7. No
8. Who is your next of kin? (Address and relationship to be stated) The answer to this question shall not be construed as in the nature of a will. — 8. Malta — 91 Ridolfo Street Sliema
9. What is your permanent address in Australia? — 9. 27 Adelaide Terrace Perth W.A.
10. Do you now belong to, or have you ever served in, His Majesty's Army, the Marines, the Militia, the Militia Reserve, the Territorial Force, Royal Navy, or Colonial Forces? If so, state which, and if not now serving, state cause of discharge — 10. No
11. Have you stated the whole, if any, of your previous service? — 11. Yes
12. Have you ever been rejected as unfit for His Majesty's Service? If so, on what grounds? — 12. No
13. (For married men, widowers with children, and soldiers who are the sole support of widowed mother) — Do you understand that no separation allowance will be issued in respect of your service beyond an amount which together with pay would reach eight shillings per day? — 13.
14. Are you prepared to undergo inoculation against small pox and enteric fever? — 14. Yes

I, Waldemar Beck do solemnly declare that the above answers made by me to the above questions are true, and I am willing and hereby voluntarily agree to serve in the Military Forces of the Commonwealth of Australia within or beyond the limits of the Commonwealth.

And I further agree to allot not less than two-fifths / three-fifths of the pay payable to me from time to time during my service for the support of my wife / wife and children.

Date 10 : 9 : 16

Signature of person enlisted.

* This clause should be struck out in the case of unmarried men or widowers without children under 18 years of age.
† Two-fifths must be allotted to the wife, and if there are children three-fifths must be allotted.

*45. Waldemar's enlistment papers*

Waldemar was assigned to the 16th Battalion as Private 6969, serving in the 23rd reinforcements. This battalion formed part of the 4th Brigade, commanded by Colonel John Monash, which had been raised in late 1914 by the Australian government with a view to strengthen the 1st Division. Monash, who had been criticised for his battlefield performance in Gallipoli, would become the Commander of the Australian Corps in 1918. Back home after the war, he became an advocate for returned soldiers and a leading figure of Melbourne's Jewish community.

The 16th Battalion had seen battle in Gallipoli and later went on to fight in France and Belgium, including at Pozières, Ypres, Polygon Wood, Hamel, Mont St. Quentin and Amiens.

The battalion's badge lies in a wreath of thistles, depicting Saint Andrew holding his cross. Its regimental motto *'Vincens'*, meaning *'conquering'*, proved to be very accurate as this was one of the most highly-decorated regiments, including three Victoria crosses. One of its Lieutenant Colonels, Harry Murray, who rose from the rank of private in only three and a half years, happened to be the most highly-decorated infantry soldier in the war.

Writing on the back of a photo of himself proudly wearing his uniform, Waldemar explained to George the colours of the shoulder patches pertaining to the different battalions:

*'Dear George,*

*What do you think of this photo? Notice the bubble on my chin. It spoils the whole effect. But as they only cost 4/- a dozen, I suppose one could not expect anything better. The colours on my shoulder are those of my battalion, the famous 16th white and dark blue. Colours on the sleeves are the distinguishing marks of each unit, for example the 11th Battalion are brown and blue and so on. Anyhow take it for what it is worth. I will send you some more if you want them. But expect to send you some more later on. Goodbye my dearest brother.'* (Letter dated 20th April 1917)

This photo had been taken at the Imperial studio in Sutton, England, where Waldemar had gone with his battalion in March 1917.

In another photo, lovingly preserved by the Gugeri family, Waldemar is wearing his uniform, holding a cigarette, standing next to Barbara who is wearing a polka dot dress, adorned with a lace collar. Probably this was one of the last photos of them together before Waldemar sailed away. Looking at the photo, one can almost sense a certain sadness in their demeanour, as they no doubt realised they would soon be separated, with no guarantee they would ever see each other again. The same feeling of desperation, to cling to every possible moment spent together, also emerges from the note quoted at the beginning of this chapter, written just a few days before Waldemar's departure.

Once Waldemar had started his training, and as his impending departure to the Front loomed, these meetings became few and far between, every single minute shared cherished and savoured forever. Another beautiful photo is a close up of Barbara wearing a wide-brimmed hat and the same elegant dress, which Waldemar might have taken to the Front to remind him of the girl he had left behind.

After enlisting on the 11th of September 1916, Waldemar started his training on the 9th of October in Blackboy Hill camp, about 25 km east of Perth, at the foot of the Darling Ranges. With around 32,000 men going through its gates for training, it is considered the cradle of the A.I.F in Western Australia. For this reason, when, in 1957, the State Government designated the area for housing, the State Executive of the Returned Services League requested a portion to be preserved as a commemorative place. Ean McDonald, a returned serviceman, designed the site in such a way that, if one stands at the entrance and looks past the flagpole, through the arches and over a stone seat, to the city of Perth, the whole line coincides exactly with sunset every 24th of April, the eve of Anzac Day.

*46. Barbara's and Waldemar's final moments together*

*47. Barbara*

*48. Waldemar with two of his battalion mates at Blackboy Hill*

A look at Waldemar's attestation papers makes very interesting reading and gives us a lot of information and insight about him. His address at the time was 357, Adelaide Terrace in Perth. As a reference, he put down the name of Sir Gerald Strickland, former Governor of Western Australia.

Waldemar produced his birth certificate and Maltese passport to prove that he was born as a British subject. A handwritten note, by Captain J.S. Duffy from the Intelligence Section Branch Staff, advised that there was *'no objection on score of nationality'*.

Regarding his employment, Waldemar stated he had done an apprenticeship as a surveyor for one year with S. Sacco in Malta, and that his last employment had been with the Government. His trade or occupation was listed as motor driver and licensed surveyor.

His medical test was conducted on the 15th September, where he was found to be fit for service. His declared age was 27 years and 6 months. Standing at 5 feet 7 inches (170 cm) and weighing 148 lbs (67 kg), Waldemar's complexion was described as fresh, with blue eyes and dark brown hair. His chest, when fully expanded, measured at 36.5 inches (92 cm). A vaccination mark from infancy on his left arm was recorded. It was also noted that Waldemar had a slight hammer toe, but this slight defect was not considered sufficient to cause rejection.

Waldemar's teeth had also been checked and it was recommended that he gets dental treatment. In fact, he had some fillings on the 6th of October. In addition, he was vaccinated on the 10th and 18th October, and again on the 29th November.

Two days before the Christmas of 1916, Waldemar sailed with his regiment from Fremantle on the *Berrima*. From the troopship, he sent a letter to his brother, dated 21st January 1917, which he posted *'from a certain place above the tropics'* a few days later.

*49. Waldemar (marked with an arrow) with his battalion on their day of departure to the Front*

*50. Waldemar*

For security reasons, any names of places mentioned were blotted out with blue or purple ink. Waldemar mentioned the unbearable hot weather which was made even worse by the crowded ship where '*we are packed like sardines*'. Due to the dangers posed by submarines, at night the troops had to sleep down below in complete darkness. Waldemar referred to the bad smell in the close living quarters, which reminded him of a sheep pen. '*Humanity can at times smell very much.*'

George would have been pleased to hear that Waldemar went to confession while on board. '*So I have been to confession at both ports of call which I consider very good indeed.*' He also mentioned them having a burial at sea after one of the soldiers from another troopship died. *'We all had to parade and as a mark of respect all the convoy stopped still for a few minutes.'*

The long sea journey proved decidedly monotonous. Wake-up call was at 6:00, breakfast at 7:15, then free time until lunch at 12:15, followed by more free time until an early dinner at 5:15. The soldiers had to undergo one hour of physical drill every day and about once a week, they were on a twenty-four hour guard. Lights out was at 9:00 in the evening:

*'I pass most of the time reading but we are all getting sick of it and I am sure that we will all be pleased when we will set foot in England.'*

Waldemar envisaged that this would be around the middle of February, when the freezing cold would be a refreshing change from the tropical heat.

He echoed the sentiments of the majority of soldiers who were itching to go to the Front:

*'...now that I am in it, I want to be at the front. Unfortunately, it will be months and months before we will be under fire at all and by that time, one does not know what might happen. In any case, I hope that no peace will be declared till we have at least had our baptism of fire.'*

His family's reaction to the news that he had enlisted was a constant worry weighing on his mind, and he was eager to receive their letters:

*'Amongst other things, there is a mailboat travelling with us so I hope to get some letters a day or two after getting to England. I am rather anxious to get some tidings of you all and I do not know how you have all taken my news. I hope and as I pray everyday well.'*

Since, apart from his brother, he had not told anyone else in Malta that he had joined up, he told George *'that if you want to keep it in the dark, you can.'* He was particularly worried about his young nephew's reaction to the news:

*'And how is Naps, mother, Vira etc?...I hope the kid is well. I wanted to send him some cards but I thought I had better not in case you do not want to tell him where I was. Anyway I suppose they will keep sending him some cards from Australia.'*

Here Waldemar was referring to Barbara and the Gugeri family:

*'By the way, I hope you have not forgotten to write to those people to thank them for all the kindness they have showed me.'*

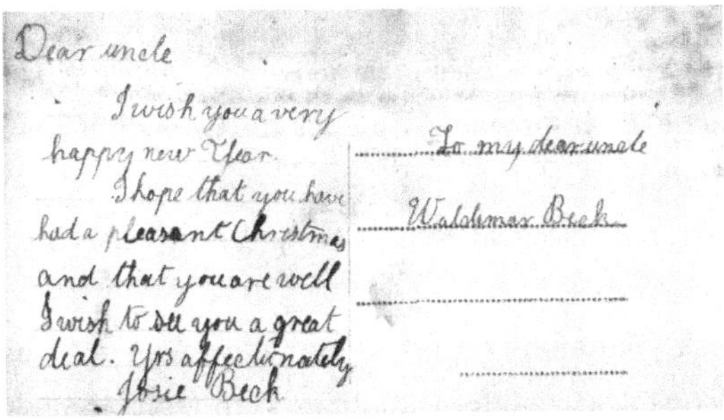

Dear uncle
I wish you a very
happy new Year.
I hope that you have
had a pleasant Christmas
and that you are well
I wish to see you a great
deal. Yrs affectionately
Josie Beck

To my dear uncle
Waldemar Beck

*51. Josie's endearing postcard to his uncle*

Before his departure, Barbara's father, Peter had given him a letter of introduction, addressed to his cousin in England:

*'My dear cousin,*

*The bearer of this is Private Waldemar Beck who is going to England having joined the Australian forces. He is a Maltese gentleman who came to Perth several years ago, and is a friend of ours - and was well known to a former Governor of this Colony Sir Gerald Strickland.*

*As he is a stranger to England, I feel sure you will be pleased to welcome him.'*

(Letter dated 6th December 1916)

Waldemar assured George that with this letter in hand, and knowing someone else from London, he should be fine once he arrived in England:

*'My chum comes from London so I have somewhere to go to when in England. Then I also have those letters of introduction so I am all right.'*

In a later letter, Waldemar was looking forward to four days' leave in London:

*'I wrote to those people, relations of the Gugeris and they all want me to go and stay with them so I am sure of a good time when I go to London.'*

The *Berrima* had called at Cape Town where Waldemar was very taken with the picturesque view of the Table Mountain. In Colombo, where the ship had previously anchored, the white population had lodged a petition to bar the Australian troops from disembarking due to previous misbehaviour. It seemed no lessons had been learnt from Egypt!

In a letter written around the same time to his friend E.J. Hayes, Waldemar referred once more to the monotony on the troopship and the immense heat of the tropics:

*'...the heat was almost unbearable. There used to be no wind, so much so that the windsails used to hang limp. Even the smoke from the funnel used to go in front, instead of trailing behind the boat.'*

By contrast, as the ship neared European shores, *'it is getting colder and colder, and we soon expect to be frozen.'*

Notwithstanding the harsh conditions on board and the rough seas, as well as the dangers from a possible encounter with *'tinned fish'* (submarines), Waldemar assured his friend that the mood among the troops was one of optimism and high spirits:

*'Still, with all that, one is not sorry. Are we downhearted? ...Most emphatically No!'*

On the 16th of February 1917, the long, tedious journey at last came to an end when Waldemar and his battalion disembarked at Davenport. Later on, he would hear the news that the troopship that had got them there, the *Berrima*, had been torpedoed in the Channel a mere day after their landing, with four of the crew killed. In a letter to George, he exclaimed: *'Our luck was in there, was it not?'*

On the 2nd of March, Waldemar wrote to George:

*'As I told you before, the voyage was very uneventful. It was a pretty long one though as we were more than 8 weeks on the water. However things went on all right with the exception of the last day when we got into a fog and a submarine must have been sighted because all of a sudden we had a lot of torpedo boats circling around us. Anyhow after a little while, things quietened down and eventually the fog lifted and we entered port.'*

Waldemar went on to write that together with the rest of his mess, he was not allowed into Salisbury at first as they had to be quarantined in a fort a few miles from Davenport, since there had been a breakout of meningitis on board. He did not seem upset that he had not joined the rest who had started their harsh training *'drilling from 7 in the morning to 7 at night.'*

At Fort Scraedon, *'we are having the time of our lives. Up at 8, bonzer breakfast at 8.30, nothing to do till lunch. After a nice march to the surrounding country, tea at 5 and generally a concert at night. We pass the days reading and playing football or else admiring the scenery from the ramparts of the fort.'*

Despite the cold weather, *'rain and fog from one day to the other'*, Waldemar still admired the beauty of the English countryside:

*'The country is very pretty, all under cultivation too. Some of the villages I have seen look very quaint. Devonport and Plymouth are very pretty ......'*

He still complained that he had not seen the sun ever since their arrival. *'Yet if we were in Australia, I suppose we would all be baked alive.'*

Waldemar wrote that, during a route march where they could see Land's End, opposite the Eddystone Lighthouse, he was very impressed watching *'five aeroplanes hunting for submarines. It was a nice experience, a sight I have never seen before.'*

Later on in France, he retained this fascination: *'There is always plenty of interest in watching aeroplanes and seeing them fight.'* (Letter to George dated 18th June 1917)

Scenes in the air like these would become more and more common. When the war started, aviation was only in its infancy; after all, it had only been ten years since the invention of the first aeroplane. At the beginning, the first war planes were very fragile and primitive, made of wood and canvas, and held together by piano wire. Pilots had to wear heavy clothing and cover their faces with whale oil to prevent frostbite, since the cockpit was open with no heating available.

Planes were first used for reconnaissance photography to monitor the movement of armies and artillery targets on the Western Front. Railways were still considered an important means of transportation during the war, as the successful and quick mobilisation of troops and supplies relied heavily on an efficient railway network. But, the idea started to gain ground that to secure an advantage in war, you had to conquer the air.

The French General Foch still belonged to the old school, believing that the army had no use for aviation. As the war wore on, he was proven wrong.

In January 1915, the English skies witnessed a strange-looking contraption: a German Zeppelin, a giant airship filled with hydrogen gas and driven by engines attached underneath. Then, in 1917, Germany developed the Gotha IV bomber with a capacity to carry a bomb load of 500 kilograms. On the 13th of June 1917, London was bombed, with 162 civilians killed, of which 16 were schoolchildren. The following year, England returned the 'compliment' by raiding Germany.

The war produced a number of legendary pilots: the French René Fonck, the British Edward Mannock, the South African Andrew Beauchamp-Proctor, and the German Erich Lowenhardt. But perhaps the most famous or infamous pilot of the time was the one and only Manfred von Richthofen, or, as he is more commonly known, the Red Baron. These were all dashing heroes whose daring aerial exploits would go down in aviation history.

Interestingly, another person by the name Waldemar Beck is found in the list of German soldiers assigned to the FFA ('*Feldflieger Abteilung*'), a field flying company. On the 10th of January 1916, while our Waldemar was still crossing the seas on his way to training in England, his German namesake was awarded the Golden military merit medal ('*Goldene Militär-Verdienst Medaille*'), of which only 4,234 were awarded in this war.

Waldemar was also greatly impressed by the ships: '*...I have seen some modern types of war vessels over here, ships I had never seen before.*'

Besides watching the skies for aeroplanes, another highlight of the day in camp was receiving letters from loved ones. Waldemar often expressed his disappointment at not hearing news from home as often as possible. If Waldemar loved to receive letters, he also enjoyed writing them. He was a prolific correspondent, writing long and descriptive letters which really captured his feelings and painted a vivid picture of his surroundings.

When I visited his family, spending afternoons reading and discussing his letters, I felt excited and privileged to have this front row glimpse into

Waldemar's life, thoughts, and opinions. The letters were well-preserved, most of them still in their original envelope, and the only sign of the ravages of time was the yellowing paper and faded ink. It was fascinating to discover and learn more about Waldemar. The missives proved a true treasure trove, from where his personality emerged through the written word, in particular his witty sense of humour, positivity, optimism, and loving and caring nature.

*'I have not heard from you since I left Australia but a friend of mine wrote from Salisbury to tell me that there are several letters for me so that I suppose I will soon be hearing from you all.'*

Waldemar was always eager to hear news of his mother and his beloved nephew Josie, or Naps. Waldemar's love for his nephew shines through in his correspondence. He was concerned about him knowing that his uncle was at war; therefore, when he sent him some cards, he addressed them to George *'so that if you do not think advisable, you need not give them to him.'*

His mother Carolina was also never far from his thoughts. He made her his next of kin *'so that if anything happens, she will get all the money.'* He requested George to ask his wife Vira that *'if later on, say after 5 months, a telegram were to come to Mum to withhold it and give it to you. You see, even if one is in hospital with ordinary sickness, the Minister for Defence sends telegrams to the next of kin and one now and again to report progress. That is the only thing I want to warn you (about). I could not manage it otherwise as I had to make Mum the next of kin.'* (Letter dated 18th of March)

He was also very interested in what was happening in Malta: *'How are things in Malta, any better?'* (Letter dated 2nd of March)

Waldemar could not hide his excitement on days when he would receive a whole package of letters, which would make up for those times when he got no mail at all. He also started marking his letters to George, to make sure he was receiving them all:

*'Whilst in the fort, I received 5 letters and a parcel from Australia but no letters from you.'*

The prospect of receiving gifts was also much anticipated, especially due to the high cost of living:

*'The price of things is something awful in England, nothing to what it is in*

*Australia, though I suppose one must expect things like this over here.'*

Things did not improve as time went by. In a letter written on the 29th of April, Waldemar again remarked that everything was getting dearer and dearer.

George had sent his brother a monetary gift for Christmas. In a letter dated 18th March, while being grateful, Waldemar assured his brother that he was financially comfortable. He also did not lack material goods and comforts, since he was well looked after by the Gugeris:

*'You can see from this that they treat us very well in Australia. I really and truly am not in need of anything at all. Why they are sending us socks, muffs and so on and galore from Guildford. In fact my kit is more than overcharged and if one of these days there will be an inspection, I am afraid that more than an article will have to go overboard.'*

*'We get cigarettes and things very cheap at the YMCA so please do not send me anything, only as many letters as you like.'*

Barbara had also sent him *'a beautiful silver locket with my regimental address on the outside and inside our mother's and her photos, really splendid.'*

Waldemar was happy that his brother had written to the Gugeris to thank them for the kindness they showed him. George had also sent them some Maltese lace:

*'They will appreciate that very much indeed. They have really been very nice to me. They are one of the best known families in WA. He is a JP and though has lost plenty of money, still they are pretty well off. He married twice. From the first wife, he had 3 girls and a son, the son is at the Front and the girls are married, one to a retiring engineer in Java and another to the magistrate of the Goldfields. By his second wife, he had 3 girls and 2 boys. Barbara is 24 years old, another one is 20 I think and one is 14. We had a lot of photos taken together and I shall tell Barbara to send you some taken at her house on my long leave.'*

Peter Gugeri had sixteen children in all but a few of them died in infancy, and this is why Waldemar only referred to the nine that survived.

In April, Waldemar wrote about the Gugeris telling him that *'they had just received a very nice letter from you and that the old chap was most highly pleased. I suppose I will hear more about it by the next mail.'* (Letter dated 29th of April)

*52. Barbara or Bobbie*

Referring to some photos and cards he sent George, *'I am glad to hear that you liked my photo. Yes everyone seems to like the one in my overcoat best.'*

Indeed, this is a great photo which shows Waldemar as a dashingly handsome young man, wearing his army coat and cap. Sadly, other photos of him taken later in the war show us a very different picture, as Waldemar, not only lost a lot of weight during his months at the Front, but also seems to have aged beyond his years.

George must have expressed his concern that Waldemar might be tempted to partake in a certain kind of 'entertainment' offered to the troops. Waldemar assured him that he needed not worry as he did not drink, gamble, or associate himself with women of ill repute. On the contrary, he was embracing his faith with renewed fervour:

*'Dear George, do not worry about me. I have altered. I went to confession both at Durban and Cape Town. I can honestly tell you that I have never drunk. As regards gambling, I cannot even play the commonest game of cards. And as far as the women, the lessons I have learnt from the venereal compound both at Blackboy and also over here, have been the biggest lessons for me.'*

He went on to say that just at Codford, there *'are hundreds with syphillis, gonorrhoea etc. And in London, some hospitals are absolutely choke full. I can give you my word of honour that I will not mix with women on any account. London is awful and our boys are suffering tremendously.'*

To put George's mind at rest, Waldemar informed him that the following Sunday, he would be going to communion and confession, *'and you can rely on me that my life, especially in that direction, will be very exemplary.'*

In later letters, every now and then, Waldemar would mention that he had been to Mass, as if to assure George that he was still keeping his promise:

*'Yesterday I went to confession and I received Holy Communion at the first Mass at 8.30. The church is at Codford about 2 miles from here, about 50 of us, officers and men received communion. I need not tell you that I prayed for you all.'* (Letter dated 22nd of April 1917)

In the same letter though, he also mentioned that, although Salisbury, located about 20 km from the training camp in Codford, was out of bounds for the troops *'still that does not prevent a lot of 'kħab'* (prostitutes) *to come at night round our camps.'*

Waldemar's only vice seems to have been smoking:

*'I have just received my weekly issue of cigarettes, four packets and a box of matches. It is a new brand of cigarettes again. It is marvellous the amount of different brands that have come out since this war started, hundreds upon hundreds of different names.'* (Letter to George dated 7th of October)

Waldemar's affairs of the soul were very well looked after by his brother and the Gugeris, who all provided him with various holy medals for protection:

*'Thank you very much for that Sacred Heart of Jesus and cross. Curiously enough, by the same mail, Barbara sent me the same identical one. They loaded me up at Guildford and I really have half a dozen medals round my neck, beads, a beautiful small prayer book from the Gugeris. So you can see how I stand. Besides, by nearly every mail, the girls send me some medal or soldier's prayer, really nearly by every mail and they tell me that old Gugeri has added a prayer for me to be said by all every night. His prayers every night are hours long.'*

Religion played an important role, with both the Allies and the Germans claiming they were the ones with God on their side. Eva Lynch, a young English girl, was quoted in Lyn Macdonald's book *'The days of hope'*, saying *'The first German prisoners were captured and they had 'Gott mit uns' inscribed on their belt buckles. We were really indignant when we heard that. What cheek! God was on our side.'*

Stories abounded among the troops of holy visions of the Virgin Mary, Joan of Arc, Saint George, or angels in the battlefields. Perhaps the most famous tale of divine intervention occurred during the battle of Mons on the 23rd of

August 1914. The odds were stacked against the British; however, miraculously, they managed to stop the Germans from advancing. Rumours spread that an army of angels came down and defended the Allies with bows and arrows. It was claimed that these were the ghosts of English archers who had died 500 years earlier in Agincourt. A British military *attaché* to Russia also reported that a wounded soldier told one of the Tsar's daughters that the Virgin Mary had appeared to the troops. Others claimed a mysterious man went around the Western Front, comforting the wounded and dying soldiers. Known as the White Helper, no bullets or shells could harm him. Many thought he was Jesus himself, since he had wounds in his hand.

In a story reported in the Maltese press on the 18th of May 1915, an Anzac was saved by a Bible in his breast pocket, and the bullet went through as far as the Gospel of Saint Matthew, marking the 43rd and 44th verse of the Second Chapter: '*Till I put thine enemies underneath thy feet.*'

Another popular belief was that dead soldiers never left the trenches, but continued to fight on with their comrades. Sceptics discredited these visions as hallucinations by battle-weary soldiers who had not consumed food or slept for days on end. Whatever the true origin of these intercessions from above, they did help to boost morale and patriotism.

On the 17th of March, Waldemar was discharged from Fort Scraedon, after being tested negative for meningitis, and transferred to Codford, where he joined his unit. Although he had previously described the fort as a '*bonzer place*', after almost a month, it had become too monotonous '*as we were not even drilled but left to do what we liked, only we had no liberty.*' Waldemar described his new place of abode as '*a big camp, full of Canadians, (illegible) and Australians.*'

The camp included training and parade grounds, lecture halls, barracks, a hospital, a cinema and Y.M.C.A recreation huts. This was the last chance for the troops to enjoy a resemblance of normality before facing the harsh realities at the Front.

Waldemar's study and knowledge of the German language served him in good stead as he was able to communicate with the German prisoners of war held at the camp:

*'There are thousands upon thousands of troops over here. Also a lot of German prisoners working in the camps. You ought to have seen their look of surprise when at times I say a few words in German to them. They look a very poor weedy lot.'*

Once again, even in the camp, receiving mail was eagerly anticipated:

*'Whilst in the fort I received 5 letters and a parcel from Australia but no letters from you. Yesterday however I got the registered letter and yours of the 4th as well as two packets of DGM. I have not had Naps's photo yet but I am sure it will soon turn up. There were also 21 letters from Australia and 3 parcels. So I was very happy indeed, really and truly happy.'* (Letter dated 18th of March 1917.)

*'Yesterday I received seven letters from Australia and yours of the 15th April. I suppose by now you have received my letter in which I told you of the safe arrival of the 2 photos.'* (Letter dated 29th of April)

*'My dear brother,*

*Last evening my luck was in as I received your letter of the 17/6 and 2 parcels from WA.'* (Letter dated 3rd of July 1917)

In his correspondence with his brother, Waldemar has left us a vivid picture of the day-to-day life in the training camp. Every morning they had physical exercises, referred to as 'jerks' by the troops. Waldemar told George that this physical activity *'also made a new man out of me.'* In fact, the long route marches helped to make the veins on his legs almost disappear.

Food at the training camp was scarce. Waldemar was concerned that if the rations were going to be cut down even further, there would be a riot. Luckily *'they give us 2 shillings a day to help us buy a few things so we can always get along but it is always a mystery how the Pommies get on.'*

Hygiene-wise, Waldemar enjoyed the luxury of hot showers and freshly laundered clothes:

*'This morning, I am stopping in at when the bathroom opens so that after tea, which on Sunday is at 4.30, I intend to have a hot shower, then a change*

*of underclothing. Afterwards, three of us are going down to the washerwoman, about 3 miles away to give her our washing. I hate washing my clothes when I can help it, though of course, once we go over to France, I will have to do it.'*

Doing his own laundry would be the least of Waldemar's worries once in France. Compared to the conditions in the trenches, where weeks would go by without the soldiers having a proper bath and a clean change of clothes, the training camp would have seemed like heaven. Despite efforts to keep a certain standard of hygiene in camp, though, thousands of men living together in close, confined quarters inevitably resulted in a fertile breeding ground for lice:

*'We have plenty of 'dud' here. Of course some of the chaps are very dirty. I know of chaps who hardly ever get a wash from one week to the other and who very seldom change their clothing and then you see, the clean fellows get them from them. We do our best such as lots of baths, constant change of underclothing and so on. We also use shirts dipped in Lysol and so on. But even with all these precautions, now and then one makes an unlucky discovery. Fortunately, I have not had them again ever since that time.'* (Letter dated 29th of April)

Here Waldemar recalled an incident at Blackboy camp before his departure to Europe:

*'Of course fellows who are very dirty receive very drastic treatment...It was about our second month in camp. There was a fellow who was very dirty...So at night, we got hold of him as he lay asleep in his singlet, it being pretty hot, ran about 100 yards with him and gave him about half a dozen duckings in one of the horse troughs.'*

About 800 people had gathered to watch this spectacle, including some officers, as such occurrences were approved, although not officially. The man in question learned a good lesson and became one of the cleanest men in camp!

The close living quarters also brought illnesses. Waldemar complained of a nasty cough, which got worse at night. The cold English weather also did not help. Despite this, he was pleased to announce to George that he had actually gained weight in camp. *'I weigh 12 stone 3, not bad eh considering that in Australia I never exceeded 11 stone 3.'* He was declared fit in all respects *'except two teeth have to come out.'* He credited this weight gain and fitness level to the route marches twice a week, as well as the daily physical drill. He was also excelling in warfare practise, undertaking musketeering and bombing courses. He seemed to have enjoyed his bayonet training, jumping over make-shift trenches and using straw figures to stab at. He was looking forward to go over to France to put all these newly-acquired skills to good use, and he felt confident about his capabilities too:

*'I think by now, I could easily hold my own against a Fritz especially if he is like some of the weedy Fritz prisoners I have seen over here. They have them over here working in gangs, putting drains, cleaning roads etc. What struck me is that they always seem to be having an easy time, lolling around, smoking and though I have watched them closely, I have never seen them very hard at work.'*

*'I think I will be getting my four days' leave at the end of this week from Friday to Monday. After that, it will not be very long before I will be going over to France...... So in about a month's time, I expect to be sent over and trained in what is known as the bullring. This is only a place where the training is continued. Of course, it will be a bit more severe but that is what is to be expected.'*

While Waldemar was still in England waiting for his turn to fight, on the 1st of April 1917, the United States entered the war. This was good news for the Allies as the Western Front would thus be reinforced by hundreds of thousands of fresh soldiers and the arrival of a much-needed supply of armaments from American factories. Speculation and talk about an American intervention had been rife since the sinking of the passenger ship *Lusitania* in 1915 by a German U-boat off the coast of Ireland, which was suspected of carrying munitions. Among the 1,200 that perished, 118 were Americans. At the time, President Woodrow Wilson had threatened that, if there would be a similar incident, the United States would consider it a planned attack.

When war broke out, the United States had remained neutral. It viewed itself as more of a mediator and committed itself to only intervene if its own interests were affected. During the 1916 presidential campaign, the Democrats' successful battle cry was that the outgoing President had kept the United States out of the war. The American public was heavily polarised about the issue, in particular the migrant population from Europe. Perhaps one of the reasons behind Wilson's reluctance to enter the war stemmed from his childhood, as his father was a chaplain in the American civil war, and the horrors of that conflict had left a lasting impression on the young Woodrow. However, in 1917, the British Secret Service intercepted a telegram from Germany to a Mexican politician, promising a slice of American territory if Mexico would join the war as a German ally. This catalyst pushed the United States to enter the war.

Another pivotal point was that in February 1917, Germany renewed its U-boat attacks on foreign ships in an attempt to block the supply chain to England. When Germany had adopted this strategy of 'sink on sight' in 1915, it only managed to destroy four percent of British ships, since the 15,000 of crossings every week were too much for the small fleet of twenty-one German U-boats. Then, this unrestricted submarine warfare adopted by the Germans aiming to attack any merchant ship, even if it belonged to a neutral country, was stopped for two years. But by 1917, Germany had increased the strength of its fleet, and by the following year, said fleet would boast almost 200 U-boats. When *HMS White Head* was sunk by a German submarine the 15th of October 1917, it was first thought that Maltese fireman Michele Falzon numbered among the dead. Masses had already been held in Malta for the repose of his soul when, to the great joy of his 'widow', news was received that he had in fact survived, and was being held as a prisoner of war by the Germans.

As the majority of ships targeted were American, the re-elected Wilson declared war on Germany, committing three million troops to join the Allies on the Western Front. Posters abounded of Uncle Sam enticing men to join. The first troops would arrive in February 1918. Although the American intervention injected a breath of much-needed fresh air to the Allied efforts, one could certainly argue that it came a little too late. By the end of the war, the US army would report 115,000 deaths and more than 205,500 wounded.

To counteract the presence of German and Austrian submarines in the Adriatic, six Royal Australian Navy destroyers arrived in Malta in August 1917, which were then deployed to the Italian port town of Brindisi. Based in Malta, they remained in service in the Mediterranean until October 1918. Spurred by this renewed German submarine warfare, Waldemar remarked *'I see in the paper that they are not going to have any more hospitals for the wounded on account of the submarining of hospital ships. I think it is a good idea.'* (Letter to George dated 22nd of April)

In the spring of 1917, another important development took place: the beginning of the German retreat behind the Hindenburg line. The latter consisted of a new line of trenches the Germans had decided to shorten to 72 km. This was far shorter than the old front established in 1914, which did not make sense anymore in view of the current situation, and also due to the overriding need to utilise troops better to free them for the Eastern Front. The Allies had to reshuffle their plans to take into consideration this new scenario. Such a development continued to reinforce the concept that this war was one of stalemate not of movement, with losses greatly disproportionate to the size of the territory gained. Moreover, as if things were not bad enough, in April 1917, nearly half of the French infantry divisions on the Western Front engaged in a mutiny. Morale was at an all-time low. The troops had little confidence in their leaders. Many fought on without any hope, and merely to avoid court marshall and execution. More than 300 British and Commonwealth soldiers were executed *'for cowardice in the face of the enemy'*. No Australians were among this number.

This was the situation that Waldemar and his battalion found themselves in when they sailed to France via Southampton on the 21st of May. On the 9th of the following month, they marched out to the unit in Étaples and seven days later, Waldemar was 'taken on strength' on the Field.

On the 22nd of May, the Base Records Office at the Victoria Barracks in Melbourne received a letter from Miss P. Douglas, who, writing from the offices of Shaw and Sons Ltd of Queen Street in Brisbane, asked for *'the official address of a friend of mine a Waldemar Beck (Maltese) who enlisted from Perth (WA) some time ago who is now in Salisbury Plains.'* In its reply, the office gave her the address as *'16th Battalion, Australian Imperial Force, Abroad'.*

On the 18th June, having some spare time, Waldemar described to George his 'home':

*'It is 3 ft high, 7 ft long and 5 ft wide, guaranteed shrapnel proof, I don't think. We were also under the impression that it was rainproof but a storm that burst over us this afternoon soon showed us that we were wrong'.*

Showing his practical side, Waldemar used his helmet to collect enough water to wash his face and hands. He shared this with *'a chap named Maskiell, a married man.' 'As a house agent would put it, it is replete with all modern conveniences and it possesses a most enchanting view.'* This to-die-for scenery included shell holes, torn barbed wire, and *'remains of what once were beautiful trees.'* In the same sarcastic tone, Waldemar added that for entertainment, they have high explosives and shrapnel, flares during the night, and gas alarms. They spent their evenings checking their underclothing for lice, which Waldemar considered the greatest nuisance they had. Food was good and *'the men are very decent and so are our officers.'*

Waldemar mentioned how he had been both in the support and front trenches, as well as involved in a 'stunt', but he did not go into much detail as *'they are neither very pleasant to write or read about.'* He just described the experience as being *'pretty tough especially the first time that we had to go through no man's land under a heavy barrage but I do not want to tell you about such things.'* To assure George, he wrote: *'Thank God I am safe and sound and I hope to continue to be so. So don't you worry about me.'*

He asked George to start sending any letters care of Mrs. M.A. Smith, at 103, Everley Road, Everley in England. Mrs. Smith was Peter Gugeri's cousin.

Writing from the frontlines on the 3rd of July, Waldemar had just got up at noon *'from a good sleep'* to write the letter, so he expressed his hope that his effort would not go unappreciated by his brother. He mentioned a group photo of officers and pipers from his battalion, which he had sent to George. He stated that he was known as Billy Beck *'for something short and rhymey',* or Wall for short. Waldemar must have felt right at home with his Australian mates and he had also started to use typical Australian slang: *'I am fast developing into a 'dinkum' soldier.'*

In the same letter, Waldemar mentioned his friend Anthony:

*'I never heard what became of Xuereb at all.'*

He must not have been the only one wondering about Anthony's whereabouts. Later in July, the Australian Red Cross also received an enquiry from someone who was preoccupied about having no news of Anthony:

*'Dear Sir,*

*We have an enquiry for the whereabouts of a man named Xuereb, a Maltese, who enlisted in Port Darwin in the first year of the War, or early in the second and who was at Gallipoli.*

*If with these meagre details you can supply us with any particulars, we shall be very grateful.'* (Letter to Base Records Office in Melbourne dated 25th of July).

The person making the enquiry gave the wrong place of enlistment as Anthony had enlisted in Brisbane, not Port Darwin, although he had been living and working there. However, as mentioned elsewhere, men from the Northern Territory had to go to Queensland to lodge their enlistment papers.

In its reply dated 1st of August, the office rectified this information and advised that *'if this soldier is identical with the subject of your enquiry, you are informed the only report received regarding him was one contained in a cable message dated London 7.12.16 to the effect that he had been wounded.'*

Anthony's movements always seem to have been elusive as the previous year, in May 1916, *'The Northern Territory Times and Gazette'* asked readers for information on Anthony's whereabouts, mentioning that he had left with the first contingent.

His service record does not have any entries for 1917, so no information exists as to Anthony's movements during that year. We know though that his battalion, the 25th, took part in the Battle of Menin Road in Belgium in September of the year in question. Earlier in May, it had a supporting role during the second Battle of Bullencourt.

After mentioning Anthony, Waldemar wrote about the Gugeri family, referring to the lace George had sent them, and mentioning how pleased they had been to receive it. He reiterated that *'they are one of the nicest families I have ever met. I am very fond of them all. Their house used to be like a second home to me.'*

George must have offered to send his brother food supplies and other comforts, but Waldemar told him not to worry as he was adequately taken care of by Barbara, who had sent him two parcels containing socks, cigarettes, small towels, soap, chocolates, Oxo, fish paste, a tin of sausages, cheese paste, paper, pencils, and envelopes. *'You can imagine how very acceptable they were to me out here.'*

The socks and towels in particular would have been very welcome and handy, especially as it had been raining *'like hell for over three days so there is nothing but slush and filthy stinking water everywhere.'* The bags that Waldemar wrapped all around his legs were of no use at all as he still got soaked through and through. He described how he had to carry out his duties at night and that was why he was still asleep at noon. He went out in a small group, crawling on his stomach, trying to keep as low as possible to avoid the *'zip zip zip of the bullets and the rat tat tat of the machine gun.'* Needless to say, the soldiers taking part in this operation *'soon become one mass of mud, clothes and at times the hands get ripped by barbed wire, but it is nothing much and all in the game.'*

At times, Waldemar's party would encounter a group of Germans doing the same job. In these instances, Waldemar proudly pointed out that *'we are pretty near always the top dogs.'*

Waldemar described the uncanny feeling he got when they used to get quite close to *'Fritz's lines.'* Being able to understand German, it would have been strange *'to be able to see and hear them quite distinctly. One wants to be more careful as at the slightest noise, a shower of bullets greets you.'*

The group would return just as dawn began to break out when they would give in their report, then *'get back to the dug out by about 3 am, have a bite, but first of all, a lovely cigarette, peel off and sleep like a top.'* Waldemar explained that the need for sleep after living on the edge: *'I suppose it is the nervous relaxation after the tension that makes one so sleepy.'*

We can imagine that, during this down time of relative calm after the huge adrenalin rush, the soldier would reflect and ponder about the close shave with being hit or even killed, and thank his lucky stars that he got to live another day.

Breakfast was served at 10:00 am and this time included bacon, *'tip top'* tea, jam, butter, cheese, and bread. *'Then perhaps if you are lucky, get a wash and smoke a few cigarettes. Then to bed again, a light lunch about 2, sleep again, tea and stew etc at 6.30. Get ready a little later on and as soon as dusk falls repeat the operations of the night before.'*

Waldemar was happy with the variety of duties he carried out. Another job which he performed involved the digging of a new trench, although this task was rife with danger, especially when one was *'standing up in the open with flares and all the infernal things flying around you. At least, it is not bad till some mate near you gets bowled over. Then you get the wind up and get pretty shaky too, let me tell you.'*

One such incident left *'poor old Nellie'*, one of his best mates, dead. Waldemar had made friends with Alfred Ernest Nelson at Blackboy Hill. On the 30th of June, Waldemar and his battalion mates had camped for the night and, in the early morning, they became the target of *'a good though a short dose of shelling full right on the dug out where poor Nelson was sleeping a few hundred yards from mine, blew poor Nellie and another chap named Marquis to smithereens and wounded the 3 other chaps.'* Nelson was described by Waldemar as a young tall chap, a good Roman Catholic *'and one of the best kind hearted fellows I have ever met.'* The only comfort was that he did not suffer any pain as he died instantly. As if he had felt a premonition regarding his fate, Nellie had given Waldemar his father's address only a few days earlier. Even though Waldemar was an avid letter writer, this surely would have been an unpleasant one to write.

Nellie had been only twenty-five at the time of his death and, in his previous life, he had been a farmer from Broome Hill. The other private killed was Clarence George Marquis, a timber-mill hand originally from South Australia. The day he died, Clarence was only twenty years old. Two cousins of his were also killed in action in the war. Like Nelson, Marquis is buried and commemorated at the Strand Military Cemetery in Ploegsteert, Belgium.

Waldemar wrote in a fatalistic tone that one does not have the luxury of thinking too much *'over these sad things which is just as well.'* He also had the unpleasant task of informing Nelson's cousin who, upon his return to the

**53. Waldemar with two of his mates in France**

trench, being one of the Brigade's runners, asked Waldemar where Nelson's dug out was. *'He got the shock of his life when I told him that poor old Nellie had met his death two days ago.'*

As time went by, Waldemar, like any other seasoned soldier, became immune to danger. He described an episode where five fair-sized shells hit too close for comfort near his dug out while he was asleep. *'But the only thing I did was to sleepily turn over and ask my mate whether the noise was from our guns or Fritz's. The only answer I got was 'Damn you and the Fritzes, go to sleep.' So there you are.'*

Despite Waldemar's apparent *nonchalance* in the face of danger, he still ended his letter with a request to George not to forget him in his prayers. Again and again, he told George not to *'worry your silly old head about me. I am allright and even a 'blighty'* (a wound) *is generally nothing much.'* Most soldiers would welcome being wounded especially if it meant being evacuated to England where, according to Waldemar, you were guaranteed *'a good OK time.'*

Waldemar appreciated the small and simple pleasures, grateful for the little blessings of trench life, such as when he inherited a dug out to himself after his companion fell into a shell hole and hurt his back. The extra space meant that *'now I can turn about and curl up when asleep. Before I had to lie with legs straight out. Now I will pull my knees right up to my chin.'*

In August 1917, Waldemar and his battalion were in Lirve where they spent twelve days in the frontline. In his letter to George dated 29th August, he expressed his excitement at sharing the yet unconfirmed news that they were going to be given two months' leave. His division was due for a spell as *'it has been on for months and months'*. Waldemar was also happy that they could ask for any supplies they needed, such as tunics, underclothing, and felt hats. This was a sure sign they would not be near the frontline since soldiers only wore steel helmets there.

When they came out of the frontline, the division's Brigadier, *'a most popular man'*, organised an impromptu canteen with beer, cigarettes, soup and cocoa. *'After a quarter of an hour stay, we lodged on again feeling very much revived.'* Undoubtedly, the Brigadier would have become even more popular after this surprise treat.

Waldemar and his division were billeted in a *'tumble down schoolhouse'*, a few miles from Nielieu for a couple of days. This was not much more comfortable than his dug out. As a result of being the target of Fritz's shots on a daily basis, *'the top rooms have all been blown to atoms and the windows are all bashed in. The floor is tiled so it was pretty cold sleeping with one blanket only.'* The schoolhouse did not offer much protection from the elements and when it started to rain, Waldemar struggled to find a dry spot. In the morning, he enjoyed the rare luxury of a bath. *'Thank God this time it was at least lukewarm not cold as ice as it was the time before.'* Afterwards, he felt human again with *'a clean change of socks, underpants, singlet, shirt and towel and at the same time the inside of our breeches and tunic were ironed.'*

Waldemar would have been interested in receiving the news that around this time, on the 3rd of August, a meeting was held on the Granaries in Floriana, Malta, addressed by the Governor Lord Metheun, who reiterated England's resolution to continue the war for freedom and justice. He mentioned Napoleon who, although a harsh leader, had never fought children, women and the elderly, which he alleged was what the Germans were doing. The Bishop of Malta, Mauro Caruana, called for the continuation of the war, which would guarantee a true and lasting peace. The bishop's stand was rather strange as it seemed to run counter to Pope Benedict XV's peace plan which had only just been issued. This plan called for a peace linked to justice rather than military conquest, a demand for a cessation of hostilities, a reduction of armaments, a guaranteed freedom of the seas, international arbitration, and the restoration of Belgium's independence.

Writing to George on the 7th of October, Waldemar referred to a letter from his brother dated 23rd September, in which George admitted that he had an attack of the blues. Waldemar almost rebuked him for feeling that way. *'You ought to be over here. I can guarantee you that you would soon get rid of all such things. You get so accustomed to dodging all sorts of things from shells to all sorts of disappointment that you soon become prone against all attack of the blues.'*

George found himself undoubtedly in an unenviable position. As the older brother and father figure, it was natural for him to worry about his younger and only brother, even though Waldemar tried his utmost to minimise the dangers of the situation he found himself in. At the same time, it would have been very difficult for George to hide his concern from his ageing mother, whom Waldemar wanted to protect from any potentially bad news.

Waldemar had kept his word in relation to matters of the soul. *'Yesterday I went to confession and this morning during the soldier's mass, I received Holy Communion. I also stopped to hear high Mass said in Flemish. It all sounded very weird. The cathedral was crowded. The congregation consisted mainly of women and old men. There were no brilliant dresses or anything of that sort. 90% of the people being dressed in crepe.'*

Waldemar's service record does not give any details as to the battles he was involved in and, for obvious reasons, he could not divulge such details in his correspondence. We can look at the movements of his battalion, the 16th, and see that, from May 1917, when Waldemar was transferred to France, to the end of that year, it took part in the battles at Polygon Wood and Ypres, spending much of 1917 in Belgium advancing to the Hindenburg line.

Notable battles that took place in 1917 include: Bullecourt from the 11th April to the 16th June and the Flanders offensive from the 7th June to the 10th November. Battles within this offensive included Messines (7th-14th June) and Nieuport (10th-11th July). The third battle of Ypres lasted from the 31st July to the 10th November. Within this battle, other battles were fought, namely

Menin Road (20th-25th September), Polygon Wood (26th September-3rd October), 1st Passchendaele (12th October) and 2nd Passchendaele (26th October-10th November). Charactarised by ceaseless rain and mud, Passchendaele was captured by the Allies but would be lost again in March 1918, and subsequently recaptured in the summer of that year.

In the first battle of Ypres in November 1914, the British, French, and German armies had lost close to a quarter million men. In 1917, another half a million were killed in the third battle. By 1918, this main city of the Flemish Westhoek would be almost annihilated. Not far from Ypres, a memorial in Ploegsteert commemorates more than 11,000 soldiers, among them the first Maltese casualty of the war - Second Lieutenant Bernard Fredrick Paul Bernard of the 2nd Battalion, Royal Warwickshire Regiment, who was killed in action on the 20th December 1914, when he was only twenty years old. He has no known grave.

Another important battle for the Allies took place between the 24th of October and the 12th of November, in Caporetto, on the Austro-Italian front. Here, Italy's army suffered great losses against the Austrians. This was the last battle for the Austrians who lost 1.2 million in the war, with Italy losing around 650,000 men.

In October 1917, an event took place in Russia which would have far-reaching effects well beyond the Russian borders, changing the course of the war. Being involved in the war from its very beginning, Russia had suffered huge losses against the German forces in Tannenberg in August and September 1914. The Russian army, although boasting millions of troops, including an all-female combat unit, was ill-supplied. A number of soldiers had to wait for their comrades to get killed to take their weapons. In 1915, the Minister of War, General Vladimir Sukhomlinov, still thought that war was best fought with sabres, bayonets and the cavalry! Tsar Nicholas II appointed himself as Commander-in-Chief and, in February 1916, ordered an attack on the Germans, which turned out to be a failure. Things continued to turn sour for Russia when, in June 1916, General Brusilov's offensive drove deep into Austria. At the same time, Romania joined the war. But when the Germans diverted troops from the Western Front to help Austria, Russia was forced to retreat. Romania was also defeated.

After the revolution, the provisional government wanted to continue Russia's involvement in the war and launched an attack on the Germans. However, the Russians were soon in retreat, and a number of them deserted. When the Bolsheviks seized power, they signed an armistice with Germany. The Treaty of Brest-Litovsk, signed on the 3rd of March 1918, granted Germany a part of Russian territory. The war on the Eastern Front was over. This was more bad news for the Allies, as now Germany mobilised its troops from the East to strengthen the Western Front.

On the 1st November 1917, shortly before they were withdrawn from the campaign in Belgium, the five Australian divisions were brought together as a single unit known as the Australian Corps. Previously, they had been under two separate units, I Anzac Corps and II Anzac Corps. The command of the Australian Corps would pass in the hands of an Australian: Lieutenant General Sir John Monash.

Yet again, 1917 proved to be a bad year for both the Allies and the Germans. The civilian populations of both Germany and England were also suffering terribly. In April 1917, Britain had only six weeks' food supply left. The situation in Germany was not much better. The Royal Navy blockade of German ports, with the participation of the Royal Australian Navy, reduced civilian rations there to only a level short of starvation. This led the German population to make bread from sawdust and potato peels, using chalk instead of flour. The previous winter, there had also been food riots in German cities.

The bad news for both sides was that, as 1917 drew to an end, there remained at least another eleven months of more starvation, suffering, and carnage.

# CHAPTER IX

# *His Life for his Country*

***'Aged 24. Another life lost. 8.8.1918. Our hearts are broken, for what?'***
*(Headstone at a Western Front cemetery)*

In his first diary entry for 1918, written on New Year's Day, Waldemar reported that *'a German plane was brought down by our chaps.'* He spent the first day of the year at Moislains, not far from Peronne, in the Somme. He also remarked that it was a very chilly day as it snowed and the wind was very cold. This reference to the icy weather is a recurrent theme in his diary:

*'Snowing again.'* (2nd January)

*'Coldest night we had so far.'* (6th January)

*'Terribly cold wind.'* (8th January)

*'Snowing all day. A brute of a day.'* (9th January)

*'Rained all day and night. Rotten weather.'* (15th January)

*'Same day: Raining whole time.'* (17th January)

As if having bad weather was not enough, Waldemar also wrote about the tedious days, and the endless boredom which was elevated occasionally by action in the skies:

*'Usual cold day. Passed as usual.'* (3rd January)

*'Usual day. Aeroplanes bombed places a few miles away.'* (4th January)

*'Fine day. Hun planes passed over camp. Nothing doing over here.'* (5th January)

Each day moulded into the next. Waldemar even forgot all about his twenty-ninth birthday on the 11th of January, which passed without any fanfare:

*'Forgot all about my birthday till the 14th.'*

He spent his birthday travelling, arriving at Bailleul at 2:00 pm, and at *'a very*

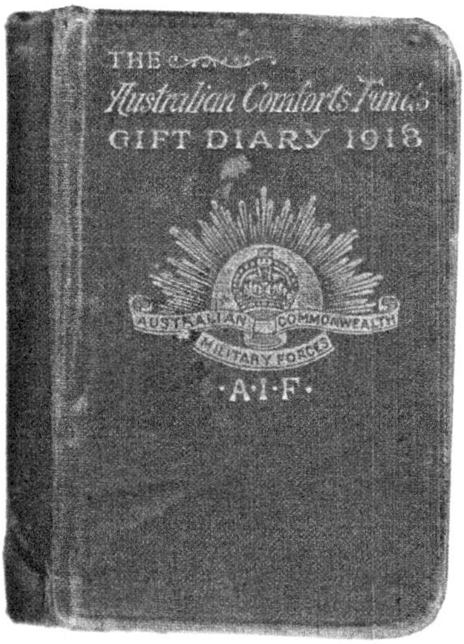


**54. *Waldemar's diary***

*decent camp'* at 5:30 pm. '*Not warm by any means*'.

The day before his birthday, though, it seems Waldemar had an early celebration as he '*got very drunk. Don't remember much. Damn all the whiskey.*'

On the 12th of January, he wrote '*usual day. Sleeping in the mess room. A bit cold. No fire but it is not too bad.*'

The next day proved to be a happy one and Waldemar could hardly contain his excitement:

'*Hurray 11 letters from Bobbie and from George. Hurray! Hurray!*'

Bobbie was Waldemar's pet name for Barbara. The previous week, on the second day of the New Year, he had wistfully written: '*Wish I could get some letters from Bobbie.*'

Throughout his diary, we encounter the same exciting tone every time the mail came in:

'*Hurray! Got 27 letters. Chaps did not tease me either! So happy. Very quiet day. Went to draw rations. Then to canteen. Going all day helping Pearce with rations.*' (30th January)

'*......Got parcel from Bobbie and cards from ?, ? Bobbie, Miss Harold and letter from Broome Hill. Hurray! Writing to Bobbie.*' (1st February)

'*......parcel from Bobbie. Hurray! Usual day. Clearer and consequently a bit livelier.*' (2nd February)

'*Hurray! 10 letters from Bobbie ...... So very glad.*' (7th February)

Waldemar liked to make lists and keep records. He used his diary to keep a tally of the letters and parcels he received. He also mentioned the postcards, photographs, and letters he sent, mainly to Barbara, his brother, and Mrs. Smith in England. He also corresponded with Mr. Gugeri, Father Raphael Pace, a Maltese priest in Perth, and Barbara's two sisters, Dorothy and Deborah, or

Doss and Debs. A Miss Harold is also mentioned as a recipient of his letters. Since the beginning of the year, he had been keeping record of his correspondence to Barbara by numbering each letter. Filling in his particulars in the diary, he gave Barbara's address in Guildford as his home address. In the next page, he wrote down the following limerick:

*'A young girl from Lahore*
*The same shape behind as before;*
*You never know where*
*To offer a chair*
*So she had to sit down on the floor.'*

The diary itself, which was issued to the troops, included handy survival hints and instructions that made trench and camp life easier:

*'Small pieces of soap which are useless for washing hands or face should not be thrown away, as further use can be found for them. They can be used for washing dishes in camp.'*

*'A bayonet makes a very acceptable toasting fork. The steel helmets are often used as wash basins.'*

A trip to Bailleul on his day off on the 18th of January had proved rather disappointing:

*'Went on leave to Bailleul, not much of a place. Pictures 'rottiest' I have ever seen – that's something. Had to walk back all the way.'*

On the 29th of January, Waldemar and his battalion were billeted at the White Chateau, which was owned by the King of Belgium. *'Not much of a place what is left of it...'* They stayed there until the 5th day of February, when they left for a camp in Dijon.

Four days later, Waldemar drew up a will in front of two witnesses Lieutenant H.R. Hillman of the 16th Battalion and A. Dowling from Kangarilla in South Australia:

*'I Waldemar Beck formerly of 91 Ridolfo St Sliema Malta regimental number 6969 now serving in the 16th Battalion of the Australian Imperial Force do hereby revoke all former wills made by me and declare this to be my last Will.*

*I devise and bequeath all my real and personal estate unto my mother Mrs Carolina Beck 91 Ridolfo St, Sliema Malta.'*

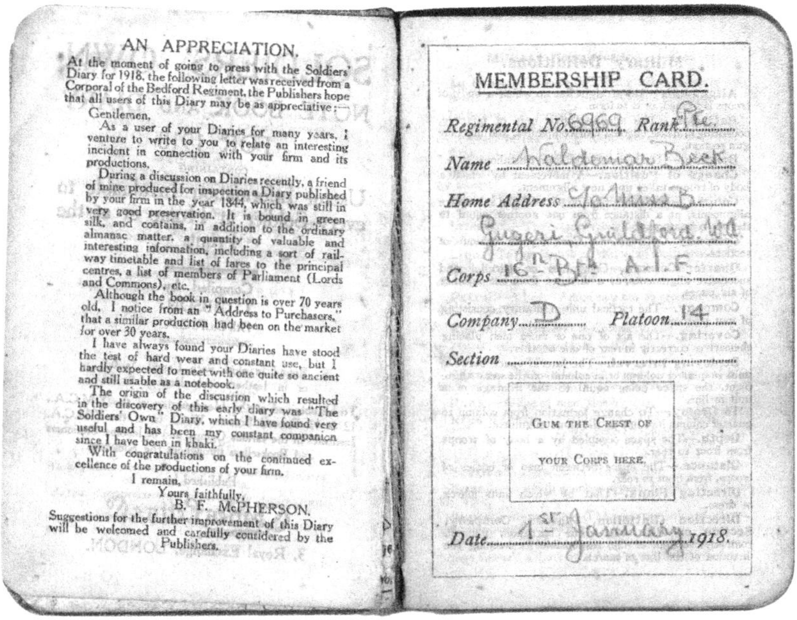

AN APPRECIATION.

At the moment of going to press with the Soldiers' Diary for 1918, the following letter was received from a Corporal of the Bedford Regiment, the Publishers hope that all users of this Diary may be as appreciative:—

Gentlemen,

As a user of your Diaries for many years, I venture to write to you to relate an interesting incident in connection with your firm and its productions.

During a discussion on Diaries recently, a friend of mine produced for inspection a Diary published by your firm in the year 1844, which was still in very good preservation. It is bound in green silk, and contains, in addition to the ordinary almanac matter, a quantity of valuable and interesting information, including a sort of railway timetable and list of fares to the principal centres, a list of members of Parliament (Lords and Commons), etc.

Although the book in question is over 70 years old, I notice from an "Address to Purchasers," that a similar production had been on the market for over 30 years.

I have always found your Diaries have stood the test of hard wear and constant use, but I hardly expected to meet with one quite so ancient and still usable as a notebook.

The origin of the discussion which resulted in the discovery of this early diary was "The Soldiers' Own" Diary, which I have found very useful and has been my constant companion since I have been in khaki.

With congratulations on the continued excellence of the productions of your firm,

I remain,
Yours faithfully,
B. F. McPHERSON.

Suggestions for the further improvement of this Diary will be welcomed and carefully considered by the Publishers.

MEMBERSHIP CARD.

Regimental No. 6969 Rank Pte.

Name Waldemar Beck

Home Address C/o Miss B.

Pugeri Guildford WA

Corps 16th Bn A.I.F.

Company D Platoon 14

Section

GUM THE CREST OF YOUR CORPS HERE.

Date 1st January 1918.

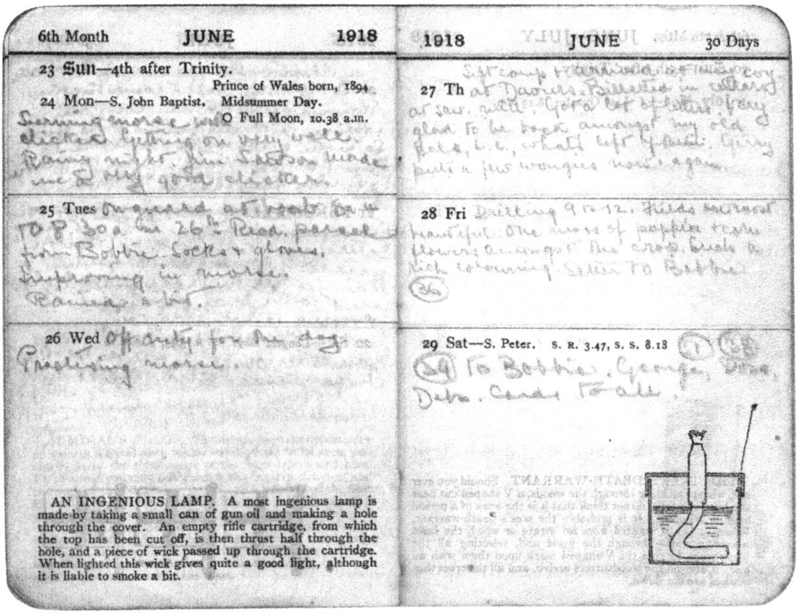

6th Month JUNE 1918

23 Sun—4th after Trinity.
Prince of Wales born, 1894

24 Mon—S. John Baptist. Midsummer Day.
O Full Moon, 10.38 a.m.

25 Tues

26 Wed Off duty for the day. Practising morse.

AN INGENIOUS LAMP. A most ingenious lamp is made by taking a small can of gun oil and making a hole through the cover. An empty rifle cartridge, from which the top has been cut off, is then thrust half through the hole, and a piece of wick passed up through the cartridge. When lighted this wick gives quite a good light, although it is liable to smoke a bit.

1918 JUNE 30 Days

27 Th

28 Fri

29 Sat—S. Peter. S. R. 3.47, S. S. 8.18

55. *Extract from Waldemar's diary*

Two days later, Waldemar had a tooth pulled out as he had been suffering from an exposed tooth nerve which had been causing him pain, especially at night.

In his diary, he also kept record of his pay in particular the times when he opted not to withdraw it:

*'Drew no pay for the 3rd time. So am in credit £4.80.'* (16th January)

*'Pay day. Drawing no money. So am in credit £5.17.4 plus £1.9.4. I have in my pocket £7.6.8.'* (31st January)

Much to the envy of their English counterparts who only received a shilling a day, an Australian private's *per diem* rate was six shillings. One shilling was put aside as deferred pay and would only be paid out upon completion of service. Soldiers then had the choice of allotting a certain amount to be left in Australia. Waldemar opted to leave three shillings at home, and received the remaining two shillings a day at the Front.

Waldemar was saving his pay as he expected to be sent on leave to England:

*'Capt pd me 60 frcs in advance. Got £8 credit in my book. Expect Blighty leave at end of this week.'* (13th February)

On the 19th of February, he was told he would be leaving for England that day. *'Biggest hurry up I have ever had.'* ... *'So happy.'*

He left Bailleul at around 11:00 am, arriving in Calais in the afternoon. Waldemar made the most of his leave in England and did not even have time to write in his diary every day. Instead, one single entry was written across the three dates from the 21st to the 23rd of February:

*'My glorious days have begun. Cheerio!'*

On the 13th of March, Waldemar wrote a detailed letter to his brother about his stay in England. He mentioned that he had *'a pretty good time during my fourteen days' leave.'* He was rather disappointed as on the days just prior to his arrival, there had been air raids but *'not a raid took place while I was in England but the day after I left another raid occurred...I wanted very much to see how the English people treat the raid.'* He went on to write that, on the other hand, he counted himself lucky, even though in France they had gotten so used to the raids, they considered them *'as a matter of course.'*

Waldemar described in detail the crossing to England. The sea was rather choppy and although the journey only took an hour, he got sea sick during the last few minutes *'when up came everything.'* By the time they arrived in London and got settled in their accommodation, it was already 3:00 am. However, he was up again at 5:30 am. *'I suppose the excitement and not being used to a soft bed prevented me from having much of a sleep.'* After spending the first day and night with some of his *'cobbers'*, the next day he went to Peter Gugeri's cousin, at whose dwelling he stayed for the next eight days. Waldemar felt at home straight away and took the opportunity to catch up on much-needed sleep:

*'I used to lie in bed till 12, have breakfast in bed and after go to town...'*

He also found time to enjoy London's vibrant cultural scene:

*'... spend the afternoon at music halls or theatres and return about 6. We used to have music and singing and go to bed about 12.'*

*'I also went to the Tower of London and to Madame Tussaud's waxworks but I could not be bothered with much sightseeing this time. I preferred concerts, theatres and such like. I suppose it was a sort of reaction after being cooped up in France.'*

In his diary, Waldemar mentioned that, on the last day of February, he went to see Oscar Asche and Lily Brayton in the musical comedy *'Chu Chin Chow'*. The next day he *'went to see Maid of the Mountains at Daly's. Bad seats...'*

Waldemar's hosts, Peter Gugeri's relatives, the Smiths, were also very musical. In fact, they *'used to be a great tenor and prima donna with the Carl Rosa Company. They used to sing in Italy, Covent Garden, New York etc and both speak Italian. Their daughter, a girl, about 28, is one of the loveliest singers I have ever heard. She also sings in theatres, concerts etc.'*

Waldemar wrote about the Smiths' sons' involvement in the war:

*'An unmarried son, very popular and a great 'dag' is in the Flying Corps. Two others are one a major and the other a veterinary captain and both are in Mesopotamia. Then there are others connected with the Daily Mail, Stock Exchange etc. The son is a mining engineer in the Straits Settlement and is married to his cousin, Mr Gugeri's daughter.'*

Here, Waldemar was referring to Peter Gugeri's daughter, Mabel, who had married Leo Smith in 1911.

Waldemar visited the Australian club, which he described as a *'magnificent building.'* He signed up with a group of around thirty Australians on leave from France who had been invited by Lord Harford for a three-day outing in the Isle of Wight. Unfortunately, the sea was very rough, so the trip had to be cancelled. Instead, Lord Harford invited them to stay at his country house near Oxford. Waldemar enjoyed this rural break and must have felt at home since, besides over a hundred deer, the manor also boasted numerous emus and a few kangaroos, as well. He also went to Oxford University and toured all the colleges.

He eventually spent his last day in London, the 5th of March, with the Smiths. In the afternoon, he was taken around the Crystal Palace by a friend of his hosts, a young priest and the padre of the Catholic sailors at the Palace. In his diary entry for the day, he wrote that he also went to confession.

Luckily, his voyage back to France was uneventful:

*'Left Victoria at 8.00 am. Left Dover 1.30 arrived at Calais 2.45. Not rough sea. Did not get sea sick. Rest camp at Calais. What a difference to the other day.'* (6th March)

All good things come to an end and Waldemar's leave must have flown by. With a hint of melancholy and dread of what he was going back to in France, Waldemar ended his letter by anticipating *'some very heavy fighting pretty soon. I do wish the damned thing would finish.'*

His last thoughts were however for the well-being of his brother:

*'I am sorry to hear that you are so overworked. It must be a damn nuisance.'*

The weather in England had been beautiful, *'nothing but sunshine and blue skies.'* But on his return to France, Waldemar had to contend with more rotten weather and rain, despite it being springtime:

*'Digging. Rotten weather.'* (29th March (Good Friday)

*'Rotten weather...'* (30th March, 4th & 6th April)

*'Letter to Bobbie (19) Raining like hell!'* (17th April)

*'Raining night and day...'* (29th April)

Waldemar rejoined his unit on the 9th of March after walking *'from Caestre to Bailleul and then to Aldershot Camp just near Neuve Eglise'* the day before. *'Pretty decent camp. Ald. Camp is at Westhof Farm.'*

His unit was the target of *'heavy shelling by Fritzy'* (14th March) and *'awful barrage by Fritzy 6 to 10.30 am. Then he attacked in 2 waves...Had plenty of firm sniping the Boches.'* (5th April)

With pride, he recorded the success of the Allied efforts:

*'Took sector held by no troops. How splendid of the Pommies.'* (27th March)

*'Got Fritzie out of Hebuterne. Still pushing him back.'* (28th March)

Despite Waldemar's optimism, in early 1918, the war seemed to be going in favour of the Germans. By March 1918, they had advanced within only 64 km of Paris.

There was more bad news for the Allies. On the 3rd of March 1918, Russia ended its involvement by signing the armistice with Germany at Brest-Litovsk, the headquarters of the German army in the East. The Germans had smuggled Vladimir Lenin, the leader of the Bolsheviks, from exile and helped him to reach Russia via Switzerland. Lenin's mantra was *'Bread to the people, land to the peasants, peace to all.'* This treaty, the first to be filmed, meant a humiliation for Russia, which had been caught up with fighting its own internal demons: the war had wreaked havoc with its economy, bringing it to the brink of financial and political chaos and disaster, amidst a bloody revolution. Russia had suffered between 900,000-2.5 million casualties, with 1.5 million wounded.

This Russian withdrawal enabled Germany to concentrate its armies on the Western Front. The Allied forces were stretched to breaking point. In a message of the 11th of April, Sir Douglas Haig, commander of the British Expeditionary Force, rallied his troops:

*'There is no course open to us but to fight it out. Every position must be held to the last man: there must be no retirement. With our backs to the wall and believing in the justice of our cause, each one of us must fight to the end.'*

Waldemar would spend only a few weeks back in the trenches as, soon after his return, he was diagnosed with scabies. This skin disease was very common due to the number of people living together in close proximity in the trenches. In another letter to George, part of which is missing, he described the fifteen hour journey on an ambulance train, which took him to a place of treatment. This letter is undated; however, a look at his service record shows he was admitted to hospital on the 19th of April. Waldemar would spend most of that and the following month being treated, rejoining his unit on the 22nd of June.

In his diary entry for the 18th of April, Waldemar wrote that the doctor recommended his evacuation because of his sores. *'However ours of course won't evacuate anyone.'*

The next day, though, he was *'evacuated at 9 am by 13th Doc at Rossignol Farm. Went by wagon to 42nd D (?) station. Evacuated by car to 3rd Field Ambulance. Evacuated by car ??? Stationary Hospital at Doullens.'*

As he explained to his brother, the treatment consisted of sulphur baths and ointments. He also gave a graphic description of the effect of scabies on his skin: *'The disease consists of little red blotches which after a while turn bluish in colour.'* He assured his family that *'there is no pain but a violent irritation which makes you scratch worse than a monkey. That is the only drawback to the main thing.'*

Once again, during trying times, Waldemar's positive nature shines true and he finds himself being thankful for small mercies. This time, he was grateful for the good food, comfortable beds and the entertainment:

*'You ought to see me in hospital. You would smile, good tucker and good beds here, pictures etc., good oh. Keep addressing letters to the same place as I am only temporarily here.'*

Even in his diary, Waldemar raved about the comfortable conditions:

*'Had bonzer feed. Stretchers and blankets to sleep on.'* (19th April)

*'Still at CCS* (Casualty Clearing Station). *Had two good feeds. Expect clearance this afternoon. Had to dump pack: equipment, rifle and blankets. Good riddance. Lost a lot of socks, souvenirs etc. Can't be helped.'* (20th April)

Waldemar told his brother not to worry about him and he also expressed confidence in how the war was going for the Allies:

*'Don't worry about me, everything is allright and even Fritzy will soon get a bally good hiding.'*

On the 22nd of April, Waldemar was transferred to the 1st Australian General Hospital in Rouen *'not a bad place'*. At the convalescent camp where he was next taken:

*56. A changed man, Waldemar in December 1916 and April 1918*

'*Had rather a bad night. Too cold. Could not get a mattress. Good tucker. Thousands in camp. Concerts (?) comfort (?) pictures etc. Got pd as I am overdrawn.*' (24th April)

There was some excitement in the camp on the following Saturday when the dressing tent caught fire. Otherwise, the days were passed '*reading, eating, sleeping. Doctor's inspection every afternoon at 2.*' On the 7th of May, he left Rouen for Le Havre where he was taken to the British Military Hospital. On the 9th of May:

'*Had all tops taken off septic sores. It did sting! Feeling allright now. Tucker good. Lecture on Papua ...also pictures.*'

'*Tops taken off and dressings. Can't sit down. Very good concert. Reading books, paper etc. Not a bad place this.*' (10th May)

'*Band concert. Same as yesterday. Lecture on India. Cigs, matches from Australia. Red Cross Fund.*' (11th May)

'*Saw doc. Temp 99.4. Abscess ordered to Hut ? as a bed patient. Had a lot of blood taken from the sores. Did not sting either!*' (14th May)

On the 12th of May, Waldemar remembered that it was Mother's Day. With regret, he wrote '*What a pity cannot write to Mum.*' Probably, Waldemar was refraining from contacting his mother as he did not want her to know where he was. In his letters to his brother, as stated previously, he had advised George that if any telegram arrived, Vira should keep it from his mother, even though it would be addressed to her as his next of kin, for he wanted to shield her from the truth.

By the middle of May, the weather had improved, although now it was '*very very hot.*' Waldemar was by now '*getting sick of lying in bed*' (17th May), and '*abscess burst during the night. Still in bed. Getting sick of it.*' (15th May)

On the 19th of May, his day was brightened when the mail came in:

'*Hurray! Letters from Bobbie and George, Mrs Smith and Miss Harold. Rcvd parcel from Bobbie.*'

On the same day, he also wrote two letters to Barbara, numbered 25 and 26, as well as letters to her sister Doss, his hostess in London Mrs. Smith, and his brother. In this letter to George, Waldemar mentioned his treatment for scabies. He had been out of action for exactly a month and had been to five different hospitals, the last two being more of a convalescent camp. He told George that he was expecting to be sent back that week, and then hoping to get a '*blighty*':

'*Fancy stopping for two/three months in a nice hospital. It will be a change not that I have reason to complain.*'

At the same time, he put George's mind at rest that he is and will be alright and asked him not to worry so much about him.

Describing his latest accommodation, he informed his brother that:

'*The last place is the best dock I have ever been in, decent beds, good tucker.*'

He spent his time reading, sleeping and eating. The entertainment included Lena Ashwell's party …'*you know the one that was for a long time at Malta.*' Lena Ashwell was a British actress and manager, known as the first to organise large-scale entertainment for troops at the Front.

Waldemar reiterated his excitement at receiving news from home and joked about having a few female pen pals:

'*Today I have been very lucky – I got no less than 14 letters, just fancy. Besides others, I know quite a number of French people, mostly mesdesmoiselles with whom I have been keeping a correspondence for several months. There I made a few girlfriends...so I am not too badly off for correspondence, am I?'*

Waldemar ended his letter by asking about his mother and sister-in-law Vira. Once again, he entreated George not to worry:

*'I have weathered it so far and I do not see any reason why I should not go on to do so. I am sure God will protect me.*

*Cheerio*

*Your loving brother Wald.'*

George would have been happy to know that even in hospital, Waldemar had not abandoned his faith. In his diary he recorded when he had confession and received Holy Communion:

*'Confession at 6'* (25th May)

*'Brown marked me for duty on Monday. Went to Confession. Fine weather.'* (1st June)

*'Went to Holy Communion. Letters to Barbara (28), George, Debs and Mrs Smith and Miss McDermott. Left Hosp and arrived at our Base.'* (2nd June)

On the 4th of June, Waldemar passed the doctor's inspection. In the evening, though, he felt very feverish, probably a side effect after being vaccinated. The next day, his arm was also very painful. *'Feel rather rotten. Asleep all the day.'*

For the next few days, Waldemar was assigned light duties. These included:

*'Guard at no.39 General Hospital where I was a patient. My post went to the Yankees. Talking to them the whole time. Decent chaps.'*

On Saturday, the 8th of June, he had some time off which he spent sightseeing in Le Havre. *'Big place. Only had 5 francs so could not do much.'*

For the next few days, Waldemar was treated for a boil on his leg:

*'Guard at 39th Gen Hosp. Treated for boil. Decent hole. Gauze shoved in hole twice a day.'* (12th June)

*'Relvd for Guard. Fine lecture on the British Navy at YMCA. Treated for boil on leg.'* (13th June)

On the 15th of June, Waldemar left Le Havre at 8:00 pm and the following day, he marched to Bethencourt, where he was billeted in a house. The weather had become rotten again, raining all night. Waldemar consoled himself with *'a few vin rouge.'* (17th June)

The next day, he left for the Front, marching for 17 km through Vignacourt, Fromelles, and Vaux. The march on the 19th of June took him through Long-champ, Cardonnette, and Allouville. Arriving at camp at 3:00 pm, the weary troops were welcomed by a *'good feed of stew and bread. First after 3 days.'*

According to his diary entry for Thursday, the 20th of June, Waldemar was located around *'Corbie, Blangy, Villers-Bretonneux ... La Motte pretty near too.'* The next few days found Waldemar being involved in some action:

*'Lobbed shells open here on two occasions. Did not get anyone. Drilling a bit. Missed afternoon parade...... Raining all the evening and night. Went to the Transport in the evening to see Brown.'* (21st June)

*'Bayonet drill, cold windy weather. Chaps getting better. Quiet night.'* (22nd June)

Waldemar also started to learn the Morse code and even practised on his day off on the 26th of June. All this practice was paying dividends as Waldemar proudly recorded that he was *'getting on very well'* (24th June) and *'improving in morse.'* (25th June). The *'very good clicker'* made for him by Jim Sateson would have also helped. (24th June)

On the 27th of June, Waldemar left the camp and was billeted in the cellars of a sawmill in Daours. On this day, he *'got a lot of letters.'* He was also very happy to be reunited with his old pals or rather *'what is left of them'*. Furthermore, they were also under German attack: *'Gerry puts a few wongies now and again.'*

The morning of the 28th of June was spent drilling. Waldemar could not help but notice the beautiful carpet of poppies and cornflowers:

*'Fields are most beautiful. One mass of poppies and cornflowers amongst the crop. Such a rich colouring.'*

Poppies would become synonymous with the war. Since the first poppy appeal in 1921, on Remembrance Day every November, poppies have been sold to collect funds to provide services for war veterans and their dependents. Inspiration for Poppy Day came from the poem *'In Flanders Fields'*, written by the Canadian doctor, John McCrae. Siegfried Sassoon was another poet produced by this era, whose writing showed the real face of the war to the world.

When he was sent back to England after being wounded, Sassoon became an avid critic of the war, accusing the powers that be that they were prolonging the war for nothing. Because of these views, he was almost court martialled, were it not for fellow poet Robert Graves' intervention. Wilfred Owen, another poet who criticised the war openly, had met Sassoon at a psychiatric hospital where he was being treated for shell shock. Sassoon inspired Owen to harness his poetic talent, which led to Owen penning unforgettable epic poems, such as *'Anthem for Doomed Youth'* and *'Dulce et Decorum Est'*. Back on the French battlefields, Owen died at twenty-five years of age, just seven days before the end of the war. His parents received the dreaded telegram on the day the Armistice was signed, at the same time the bells were ringing, announcing the end of the war.

On the 29th of June, Waldemar recorded in his diary that he wrote three letters to Bobbie and more to his brother and Barbara's sisters *'Doss and Debs. Cards to all.'*

A few days later, Waldemar's brigade, the 4th, together with the 6th and 11th brigades, took part in the battle of Hamel. On the 4th of July, these three Australian Brigades advanced 2 km, taking the village of Hamel, the German positions on the ridge beyond, enemy equipment and over 1,600 prisoners. The infantry was assisted by sixty tanks of the British 5th Tank Brigade, breaking down wire obstacles and dealing with the enemy machine gun positions. Orchestrated by the new Australian Corps Commander General Monash, the Allied success in this battle was the result of the combination of the use of tanks, artillery, aircraft, and infantry. Hamel was also important as four companies from the U.S. Army 33rd Division fought alongside the Australians, this being the first time that soldiers of these two nations fought side by side.

Lasting only ninety-three minutes, the battle of Hamel was a small battle in comparison with other more famous confrontations. However, it was a huge success for the Allies with relatively few casualties.

Sadly, for Waldemar, he was one of these *'few casualties.'* These words naturally would have been of no consolation to his family and loved ones. His promising young life of just 29 years and 6 months had started in Malta, took him all the way to Australia, and then ended abruptly in battle-weary France.

In a letter to George dated 26th of August, written by Frank Ritchie Bullen, we have a poignant firsthand account of Waldemar's last moments:

*'When the battalion moved into the frontline, he became personal runner to the company commander at that time Captain Lynas. Private Beck was wounded in the head by shrapnel when taking part in a raid on the enemy's trenches just before our big attack on Vaire Wood south of the village of Hamel. He did not appear to be badly hit as he walked by himself to the dressing station. The wound must have been more serious than it appeared as within a week or so he was reported as having 'died of wounds.''*

Describing himself as *'one of his many friends'*, Bullen was full of praise for Waldemar, not only for his friendly character but also for being an exemplary soldier:

*'He was most popular in the company. For some considerable time, his duties out of the line consisted in helping in the Officers' Mess or in other words he was practically Officers' Mess caterer.*

*He was a general favourite among his chums and moreover always did his duty. The OC officers and remainder of the company join with me in expressing their deep regret at his decease.'*

Bullen, who had enlisted on the 24th of November 1914, survived the war and returned to Australia in December 1918.

Waldemar's service record gives more detail about his cause of death. He was wounded in action on the 4th of July, with a gunshot wound through his right eye that penetrated his brain. He died two days later at the 8th General Hospital.

In the next few days, the army bureaucracy followed its usual procedures following the death of a soldier. A copy of his statement of service was sent to the Pensions Office. His will was located and the Office of the Public Curator in Perth was contacted.

We can only imagine how heart breaking it must have been for George to

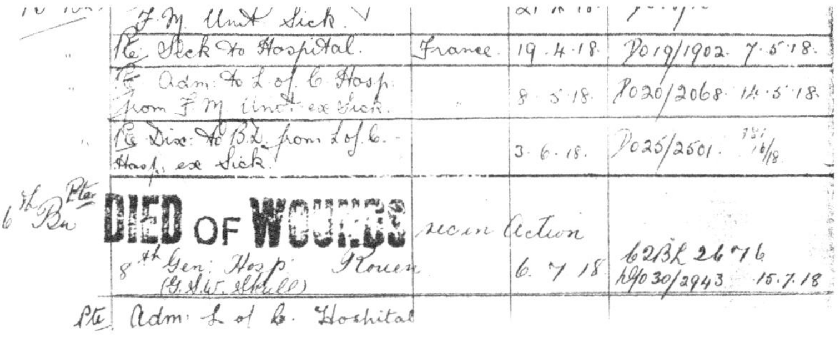

| | | | |
|---|---|---|---|
| F.M. Unit Sick. | | | |
| Re Sick to Hospital. | France. | 19.4.18. | D019/1902. 7.5.18. |
| Re Adm: to L. of C. Hosp. from F.M. Unit ex Sick. | | 8.5.18. | D020/2068. 14.5.18. |
| Re Disc: to B.D. from L of C. Hosp. ex Sick. | | 3.6.18. | D025/2501. 16/6/18 |
| 16th Bn Pte DIED OF WOUNDS | recd in | Action | |
| 8th Gen: Hosp Rouen (G.S.W. Skull) | | 6.7.18 | 62BR 2676 A/D 30/2943 15.7.18 |
| Pte Adm. L of C. Hospital | | | |

*57. Waldemar's record stamped 'Died of wounds'*

receive the much-dreaded official letter advising him of his young brother's death. It must have been even harder for him to break this terrible news to his elderly mother. And in addition, it would also have been extremely difficult for Waldemar's twelve-year old nephew Josie to come to grips with the fact that he would never see his uncle again. At the back of a photograph of a battle-weary Waldemar dated 10th April 1918, Josie would years later type the following note:

*'Uncle Waldemar (Ziju Voli) taken less than 3 months before he was killed in action 6/7/1918). His very last known photo.'*

Josie would later call his only son after his beloved Uncle Waldemar.

Across the seas, thousands of miles away, Barbara and her family would have experienced the same deep sorrow and grief. Barbara published a number of death notices for Waldemar in different newspapers:

*'Pte. Waldemar Beck died of wounds on July 6 1918 in France after 13 months in the trenches. Beloved second son of Mrs C. Beck, Malta. RIP. Inserted by Barbara Gugeri, Victoria Street, Guildford.'* ('The West Australian', 27th of July 1918)

*'Pte. Waldemar Beck who died of wounds in France on July 16* (should read July 6), *1918. Pte. Beck, who was a native of Malta, where his mother still resides, was a University graduate. He came to Westralia about four years ago. He made a number of friends in Perth, and was highly respected.'* ('The Swan Express', 9th of August 1918)

Another notice appeared in *'The Western Mail'* of the 2nd of August 1918.

Decades later, Barbara's nieces also felt for their aunt's loss:

**DIED OF WOUNDS.**

**PTE. WALDEMAR BECK,**

16th Battalion, who died of wounds in France on July 16, 1918. Pte. Beck, who was a native of Malta, where his mother still resides, was a University graduate. He came to Westralia about four years ago. He made a number of friends in Perth, and was highly respected.

**'W.A. Record" block.**

*58. A tribute to Waldemar in 'The Swan Express'*

'*......the love affair did sound very romantic didn't it? Very sad though to end the way it did.*' (Email from Debbie, 17th of October 2001)

'*It* (Waldemar's diary) *is interesting and very sad as the entries suddenly stop.*'

'*I must say that I feel terribly sad when I read all the war records. I almost feel now that I knew him – he has become so real and really a rather important part of our family history...*' (Email from Barbara, 18th of October 2001)

Among the photos of Waldemar passed onto her by her aunt, Barbara has a memorial card with a photo of Waldemar in his uniform. The card reads:

*'Of your charity*
*pray for the soul of Waldemar Beck, B.E. & A., L.S. & A.*
*16th Battalion, 4th Brigade*
*Australian Imperial Force*
*who died of wounds*
*at the British Western Front*
*on the 6th July 1918*
*at the age of 29 years.*
*R.I.P.*
*Thy will be done.*
*His life for his country*
*His Soul to God'*

Waldemar was buried in the Boisguillaume communal cemetery extension,

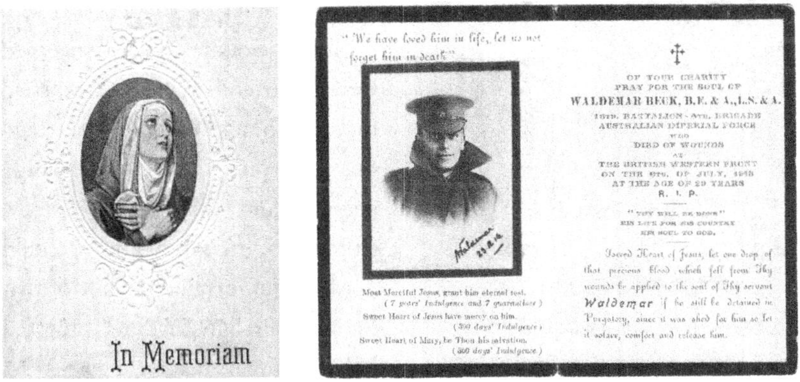


*59. Waldmar's memorial card*

at Seine-Maritime in France, about 5 km north-east of Rouen on the main road to Neufchatel. On the 2nd of July 1920, two photographs of the grave number 6B located at Row D were sent to Waldemar's mother. The booklet *'Where Australians rest'* was sent later, containing information on the gravesites:

*'Dear Madam,*

*Forwarded herewith are two photographs and description of the locality of the grave of your son, the late No.6969, Private W. Beck, 16th Battalion, which I trust come safely to hand. Any discrepancy appearing on the temporary memorial will be rectified when the permanent headstone is erected.*

*Yours faithfully,*

*Major*

*Officer i/c Base records.'*

A very touching letter, kept in the Beck family collection, was sent to George from a gentleman who visited Waldemar's grave on the 25th September of the year 1934. Three photographs accompanied the letter, including a close up of the headstone and one of the gentleman standing respectfully next to the grave.

*'Dear Mr Beck,*

*I hasten to give you a report of my visit to the Boisguillaume cemetery as I have no doubt that you are very anxious to know whether and in what manner I have carried out my promise to visit your brother's grave.'*

***60. Graves of the fallen pamphlet***

The weather on the day was very cold and cloudy, which concerned George's friend as he was worried that any photos he would take would not turn out. The country setting engulfed him with a sense of peace, although at the same time, he could not help but imagine the horror that the same surroundings had witnessed only a few years before:

*'From Rouen one takes a tram to Boisguillaume, a quite suburb hidden away in trees almost surrounded by fields. Peace reigned everywhere and as I mended my way into one of those French country lanes we know so well from war time, photos of the turmoil of 20 years ago came back to my mind and I saw a vision of German troops advancing, of the cruel and terrific bombardments which laid waste and barren the very field which I was crossing and of columns upon columns of Tommys advancing cheerily to stamp the Hun to the tune of 'It's a long way to Tipperary.'*

*'...Peace, peace and quiet and verdant pastures, fertile fields as far as the eye could see.'*

Upon his arrival in the cemetery, he came across a gardener who took him to the room where the register was kept. In an index issued by the Imperial War Graves Commission, he found the entry related to Waldemar:

*'Beck Private Waldemar 6969*

*16th Battalion Australian Infantry*

*Died of wounds 6th July 1918*

*Age 29*

*Son of Lawrence Edward and Carolina Beck of 91 Strada Ridolfo, Sliema Malta*

*D6B'*

Then the gardener and the gentleman proceeded to the grave site. George was assured that the cemetery was well-kept:

*'I found the grave, as were all other graves tended with a care which partook*

*of a labour of love. Flowers grew all around and not the least sign could be seen of any weeds at all.'*

The day had been cloudy and grey, but as the man bowed his head in silent prayer, *'the sun shone forth for a while and its light coming from heaven in glorious streaks brought to my mind those great words which are the greatest consolation ever vouchsafed to mankind: 'I am the light, whoever believeth in me shall not die forever.'*

After taking a few photographs, the man asked the gardener to put some flowers on the grave, and before leaving, he signed the visitors' book. Reading the previous entries, he was considerably touched by some of them. He looked around him and feeling the peace *'which they had given us through their death'*, he was inspired to write:

*'May their sacrifice bear fruit in peace and good will among men. Amen.'*

As he made his way back across the fields, he found it hard to describe exactly how he felt: *'Thoughts and memories rushed through my mind but out of chaos, I was aware of a concrete formulation of one wish, one ideal to cherish and strive for. Let us have peace and not break faith with those who lie in Flanders fields.'*

***61. George's friend paying his respects at Waldemar's grave***

Unfortunately, at the time this letter was written, in the early thirties, dark sinister clouds were once again gathering in Europe, which would cause another war, even deadlier than the first.

Waldemar is mentioned in the book about the 16th Battalion, *'The Old 16th'* by C. Longmore, in the list of the members who lost their life in battle. As Paul Calleja observed in his book *'Maltese of the Western Third'*, Waldemar was *'killed only a few months before the war's end. Whereas his friend Charles Bonavia became one of the first statistics of the war, Beck became one of the last.'*

As for Anthony, he may not have been aware of the cruel fate his friends had met. His movements around this time took him on leave to England on the 16th of August, rejoining his unit on the 4th of September. Anthony's battalion, the 25th, was heavily involved in the major battles of 1918, and particularly in its attempt to push the German spring offensive in March and April. Waldemar's battalion, the 16th, had also been involved in the spring attack of German positions, and in April at Bullecourt, together with the rest of the 4th Brigade, it had suffered heavily after the promised tank support did not materialise. Furthermore, at this stage of the war, Germany was strengthened with the added manpower that had been relieved from the Eastern Front.

'*The Northern Territory Times and Gazette*' of the 13th of July included a contribution by Leslie Giles, a fellow 25th Battalion member who, under the heading '*Our Boys at the Front*', wrote:

*'We had quite a reunion at a certain part of the line and we made a brave attempt to get a photo group taken but alas the weather was against us for after walking around three km through muddy ways from various map references, the weather was too dull and the photo a failure. Among those who turned up were Dave and Harold Whiteford...Xuereb and the writer.'*

The last battle Anthony's battalion was involved in, before being disbanded, happened around Beaurevoir on the 3rd of October. Earlier attempts in September to disband the 25th and merge it with other battalions had been resisted by its members. They asked to be given the most dangerous duties in the subsequent battle as they argued that, after fighting such a battle, there would either be no battalion left, or it would become so legendary that no one would dare to disband it. On the 4th of October, Anthony was injured on the Field after spraining his right ankle. A day later, he was admitted to hospital, and on the 7th of October, he was transferred to the 4th Southern General

Hospital in Plymouth, England where he stayed for twenty-eight days. On the 4th of November, he was admitted to the 4th General Hospital in Devonport after being taken sick. Two days later, he was examined and his ankle was said to have *'some swelling'* making Anthony experience *'pain in walking'*. On the 26th of November, he was moved to the barracks in Tidworth.

With his battalion, Anthony would also have been involved in the attack of the 8th of August 1918, the Battle of Amiens, considered to be a turning point at this advanced stage of the war. This battle followed the plan used at Hamel, which had proved such a success. Aided by the Canadian Corps, the five Australian divisions took almost 8,000 prisoners, over 170 guns, and enough engineering material to last them the rest of the war, making this battle one of the strongest blows against the German line. No wonder this date was labelled *'der schwarze Tag'*, (*'the Black Day'*), by the Germans who, after this fiasco, started to seriously consider an armistice.

Things had not been going well for Germany. The Allied blockade of its ports impeded important vital supplies to reach its population. The railway network was collapsing due to lack of maintenance and food shortage. Strikes and mutinies were the order of the day. The Germans were pushed further and further back towards the Hindenburg Line.

Ironically, during the last weeks of a war that had dragged on for four years, things moved very quickly. By September of 1918, Germany's ally, Turkey, had been brought to its knees in the Middle East by the combined might of the British troops, the ANZAC mounted division and the Arabs led by T.E. Lawrence, the notorious Lawrence of Arabia. On the 30th of October 1918, Turkey signed the Armistice, and was divided by the Allies between Britain, France, Italy, and Greece. The sun had finally set on the Ottoman Empire. Germany's two other allies, Bulgaria and Austria, also collapsed. On the 7th of November 1918, a German delegation crossed the frontline to discuss peace terms with the Allies. Three days later, the Kaiser abdicated and escaped to Holland.

On the 11th November 1918, Germany signed the Armistice in a railway carriage in the French forest of Compiègne. Adolf Hitler later would denounce those Germans who signed the Armistice as *'the November Criminals'*.

Twenty-two years later, in the middle of an even deadlier world war, he would force the French to sign another armistice in the same train carriage.

The guns fell silent on the same day at 11:00 am. At home, jubilant crowds gathered in the main streets of cities and towns, almost in disbelief that a war that had dragged on for four years was finally over. Mixed emotions reigned among the war-weary troops:

*'Eleven o'clock on Monday the 11th of November 1918 is, I suppose, the greatest hour of the greatest day the world has known and it has been gone just an hour. Ps And I am still alive.'* (Corporal Edgar Morrow, 28th Battalion, writing to his family.)

News of the Armistice caught other soldiers by surprise. For others, the news that the war was over provided an anticlimax of sorts. In a way, they were scared to go back to their old lives as they did not know what they would find upon their return home: perhaps their loved ones were now lost to them after all the waiting, they had no jobs to go to, and had little in common with the family and friends they had left behind. They would miss the camaraderie and strong friendships they had forged in the most dangerous and challenging circumstances. Also, they felt guilty in leaving their dead mates behind.

Private J.P. Scott of the 45th Battalion echoed these mixed feelings:

*'One is jolly glad to be out of it, yet...men you have been friendly with and stood side by side for months or perhaps into years...have been killed – one's heart fills with sadness – and one has a hankering to be back over there with 'the boys' once more.'*

Australia at the time only had a population of around five million people. 416,809 had enlisted, representing 38.7% of the male population aged 18-44, with more than 330,000 serving in distant battlefields in Europe and the Middle East. Australian battle casualties totalled 215,045, a percentage of 64.8 of the troops in the field. 60,000 of them never made it back home, of which 46,000 died on the Western Front. In Gallipoli alone, Australia and New Zealand lost 8,000 and 2,700 of their sons respectively, with another 58,000 wounded.

Other countries on both sides of the divide fared even worse. In total, more than 9 million military personnel, comprising more than 5 million Allied troops, and a further 3.5 million from Germany and the Central Powers, as well as around 6.5 million civilians were dead. Many more millions were maimed for life, both physically and psychologically. Although statistics vary depending on which sources are consulted, one cannot deny the fact that a whole generation of young men had almost been exterminated. It was indeed a 'world war' as it had spread its deadly tentacles around the globe: from Turkey to France, all the way to Papua New Guinea, to Sudan, Iraq, Africa, and as far as the north-west frontier of India.

The Christmas of 1918 was the first 'peaceful' Christmas since 1914 - a time that brought forth a whole array of mixed emotions. For those who had suffered the loss of a loved one, like Charles's and Waldemar's families, both the 'victory' celebrations and the yuletide festivities would have seemed empty and meaningless. For the lucky ones, like Anthony, who survived the war and had started to return home and be reunited with their families, their joy would have been tinged with sadness as the reality of what they had been through sank in. They would never be the same again. Few were prepared for the hard task ahead of settling back into a civilian existence and rebuilding a new life. The war may have been over, but their very own internal battle had only just begun.

# CHAPTER X

# *The Aftermath*

***'They say it is a wide road that leads to war and only a narrow path that leads home again.'***

*(A young recruit from Kiev)*

The Armistice was followed by a peace agreement, the Treaty of Versailles, signed in the majestic Hall of Mirrors on the 28th of June 1919. This happened to be the fifth anniversary of the assassination of Archduke Franz Ferdinand, which had triggered the chain of events that led to the commencement of hostilities. Interestingly, the number plate of the car he was assassinated in was A 111 118. Many pointed out that the 'A' stood for 'Armistice' and the number depicted the exact date when it was signed: the 11th day of the 11th month of the year 1918.

The Agreement seemed to create more problems than it sought to resolve. Germany considered the conditions imposed on it to be too harsh. These included reparation payments, the surrender of its armaments and fleet, and the return of certain territories. The only advantage for Germany was that the treaty brought to an end the Allied blockade of its ports that, even in the year 1919, was still causing millions of German civilians to suffer greatly from sickness and starvation.

The head of the German delegation, Ulrich von Brockdorf-Rantzau, remarked sarcastically that the treaty should just have one sentence: '*Germany renounces its own existence.*' Politician Matthias Erzberger, who signed on behalf of Germany, concurred: '*A country of 70 million people is suffering but it will not perish*', to which the French representative Ferdinand Foch curtly replied: '*Très bien.*'

Despite the German misapprehensions, the agreement was not strong enough to weaken Germany. The main reason was that the Allies all reacted differently to the treaty. While France took a stricter view in regards to its enforcement, Britain was more conciliatory. The United States rejected the treaty and negotiated separate terms with Germany. This disjointed approach failed to curb German power and control the rise of nationalism. Despite the efforts of the League of Nations, set up in 1920 to prevent another global conflict, these new instabilities paved the way for another war in twenty years' time, when the world was still recovering from its deep wounds and scars caused by the First World War. This war had reached a global scale like no other before it, creating a whole generation of youth in disarray, and leaving millions of families decimated by the destruction and deaths of both civilian and military populations. Vera Brittain, a British nurse who had lost her fiancé, brother and several friends, felt that the treaty:

*'...did not represent the kind of 'victory' that the young men whom I had loved would have regarded as sufficient justification for their lost lives.'*

The First World War was the result of the need for revenge and restoration of national pride after humiliating defeats in previous conflicts, such as the Franco-Prussian War of 1870-1871 and the Balkan Wars of 1912-1913. Likewise, the Second World War resulted from unresolved issues and built-up resentment, the direct consequences of the long-term repercussions created by the first one. The war that was to end all wars had practically solved nothing. If all wars are futile, then this one had been the most futile of them all.

With the extermination of the Ottoman, German and Austro-Hungarian empires, the borders of Europe had to be redesigned, and new maps redrawn. Old world countries collapsed and new countries such as Yugoslavia and Czechoslovakia emerged. The sun had also set on the Russian Empire when Tsar Nicholas II abdicated in March 1917 and was later executed with all his family. On the 21st of April 1919, his mother Maria Feodorovna arrived in Malta, with her daughter Grand Duchess Xenia and her five children amongst others. She stayed at San Anton Palace for nine days and then left for London to live with her sister Alexandra, widow of King Edward the VII.

During the revolution and civil war that ensued, about 800 refugees came to

Malta from Russia. Some of these were seen selling off their jewellery in Strada Rjali, the main thoroughfare in Valletta. Many Maltese had also lost a lot of money from investments in Russia. The situation in Malta was very precarious after the war due to the scarcity of basic foods and a high increase in the cost of living. A series of riots against profiteering merchants and the British colonial government culminated in the *'Sette Giugno'* uprising on the 7th of June 1919, when the British troops fired into the protesting crowd, killing four men. This event is seen as the first step in Malta's path towards self-determination and independence.

The war had also touched the Middle East and created the Palestinian issue, which would continue to haunt the international scene for decades to come. Germany lost Cameroon, Togo, German South West Africa, and its colonies in the Pacific. The United States, on the other hand, was establishing itself as a strong nation and a force to be reckoned with. Similarly, Australia also changed the way it viewed itself and the rest of the world. It had proven that it could stand on its own two feet in the world arena, emerging as a nation in its own right, no longer in the shadow of the Empire.

Although on the losing side, Turkey would start its journey into the modern world after a programme of reforms introduced in the 1920's by Mustafa Kemal, the Turkish general who had fought so gallantly and bravely for his country. Under the Armistice, Britain, France and Italy had been given control of Constantinople. Kemal led the fight against this Allied occupation, which culminated in the 1923 War of Independence at Izmir, or, as it was known at the time, Smyrna. When Atatürk, as he became known (meaning *'Father of the Turks'*), abolished the Sultanate in 1922, Sultan Mehmed VI was exiled for some time in Malta.

If anyone thought that the war finished once and for all after the Armistice was signed, they were gravely mistaken. Its effects were long felt many years later. The countries on the so-called winning side had to deal with crippling economies. The cessation of trade due to war led to unemployment and bankruptcy, and would eventually be one of the causes of the Great Depression.

Then one had the logistical problems of repatriating soldiers and their integration back to civil society. The returned men worried about their future as, after having spent years outside the workforce, some of their skills were no longer recognised and this meant they had to start from scratch. Even if they were offered their previous employment, some felt that, after what they had been through, their war service was being devalued and their sacrifice unappreciated. Men who were once strong and able-bodied came back maimed for life. They were let down by the slow wheels of bureaucracy that caused delays in receiving much-needed financial aid, such as disability pensions. Financial difficulties led some to become beggars. General Douglas Haig's future, on the other hand, was guaranteed to be a comfortable one after he was given the princely sum of £100,000 by the British government.

In 1919, the major cities witnessed a wave of violence and drunkenness as men struggled to fit back in. They had gotten used to the routine of war, which, in a strange way, had given them security and a purpose. Ironically, in certain respects, they found it harder to live in peace than in war.

Worse than the physical or economic effects were the invisible wounds. The First World War was the first conflict that made use of the inventions developed during the Industrial Revolution for destructive purposes, which left unprecedented carnage in their wake. The great developments in technology and communication made it possible for the war to be fought simultaneously on three different fronts: land, air and sea. On land, tanks were used to great effect. For the first time in history, civilians were exposed to air attacks, and at sea, submarine warfare was developed. The use of poisonous gas and more effective armaments caused unprecedented suffering and death, and the resulting physical and psychological injuries continued to affect people long after the last shots were heard.

In Australia alone, 2,000 returned men spent the rest of their life in hospital. 22,742 veterans were in hospital in 1926, and in 1939, twenty-one years after the war had officially ended and the year another one was about to start, there were 49,157 hospitalised veterans. Shell shock or, as it is now called, post-traumatic stress disorder, caused these men to sob and shake for no apparent reason. They walked with a distant look in their eyes, oblivious to their surroundings, and unable to recognise their own family. Recalling the loud explosions in the trenches, any noise frightened them and they were plagued by sleepless nights and bouts of despair. The emotional shock and nervous exhaustion would haunt them for years, and sadly for some, these symptoms would remain a constant for the rest of their life. Dorothy, one of Barbara's sisters, who was a matron, worked with the Rehabilitation Department where she lovingly nursed a number of these men.

Those who returned from the battlefields came back with their lives changed forever. Although their perception of the world had altered after the horrible events they had witnessed, they were pressured to fit back into society as if the war had never happened. Some relatives and society at large expected the returned soldiers to settle down with ease, and the warm and jubilant welcomes soon gave way to high expectations. The world wanted to move forward and forget, while the returned soldiers were coming to grips with what they had seen and still processing it. They felt isolated and misunderstood from the rest of the community, and thus, became strangers to their own families with nothing in common with them. Sometimes they found that although their families meant well, those people could not understand the trauma and pain that still burdened them long after the war had finished. They needed all the support they could get to reintegrate themselves into society. However, with no formal counselling programme in place, it was hard for them to speak about their experience, or they only opened up with their mates. This was a time when one was discouraged from talking about and working through one's feelings. It was believed that big men did not cry and the notion of the sensitive new age man was still decades away. Some took to drink; others went mad or killed themselves. Everyone was expected to 'soldier on'. After all, the worst was over, wasn't it?

In an edition of the English magazine '*Women's Weekly*', a returned soldier made the following heartfelt plea to the readers, just five days after the Armistice:

'*You people at home cannot, by the wildest stretches of the imagination, know the effect of the sudden change from the life out there to the inside of an English home. It is all much too sudden, we are not prepared for it. We ought to have at least a month's leave. The first week we should spend chiefly in our own company, or with civilian men. Do you realise what it means to go for months at a time without speaking to a woman? To have nothing but men around you night and day?*

*......You people at home look upon me as the same Jack who went away four years ago, and I am in the sense that I love you all just as dearly as ever, if not more, but in every other respect, I am a different chap. My nerves are all jagged. When one has seen the sights and the sounds I have, how can you expect them to be otherwise?'*

Although the extent of the butchery and bloodshed had never been witnessed before, all was not lost for humankind. The courage and tenacity that emerged from the worst circumstances would continue to inspire generations to come. As the official war historian, Charles Bean rightly put it:

*'The spirit of Anzac stands for reckless valour in a good cause, for enterprise, resourcefulness, fidelity, comradeship and endurance that will never own defeat.'*

In Australia, the Returned Services League, Legacy, and other Ex Services associations strive to this day to keep this Anzac spirit alive by promoting the values of mateship, valour, humour, integrity, dignity and patriotism. One cannot mention enough the great contribution these, and other similar organisations around the world, have made over the years, in providing a haven where people who had passed through the same traumatic experiences can get together and share their stories.

The returned men had changed, but the world they came back to was also unrecognisable in certain aspects. Old values disappeared and made way for new standards and ways of life. Perhaps the most astounding change occurred in the way the war contributed to the emancipation of women. While their men were away, the women had carved a new role for themselves and upon their return, a number of the men could not recognise their own wife or sister. Some wives and girlfriends turned cold as they had been sick of waiting for their men to return. The wives had been expected to keep the home fires burning for years on end, while raising a brood of children on their own. Others could not accept that their once-strong, independent man was now an amputee, shell shocked, gassed, blinded, or ill.

During the war years, women proved their worth, taking over the jobs vacated by men who were fighting at the Front. They worked long hours in ammunition factories and toiled in the fields. As nurses, they risked their lives helping and caring for the wounded on the frontlines. The Russian army even boasted female snipers. After the war, the women did not go back to their kitchens but remained in their jobs. There was no turning back. When the war broke out, the Suffragettes had suspended their civil disobedience campaign to raise awareness about extending the right to vote to women, and instead they pledged their support to the war. In 1918, British women over thirty were finally granted the right to vote and stand for elections. Australia and New Zealand were more *avant-garde* in this respect as women's suffrage had been introduced in 1895, (in the state of South Australia), and in 1893 respectively.

Women started to find their place in the world as equal partners and active contributors in the work force and community, and began to enjoy a new-found freedom. Chaperones disappeared and more women started to wear cosmetics and smoke openly. Their hair became shorter and their clothes less restrictive, anticipating the liberal fashions and views of the 1920's. The social fabric was changed forever and some returned men found this hard to accept.

*62. Anthony's visit to Malta after the war*

This more open and liberal view of life is reflected in a photograph taken in Malta after the war, showing Anthony in uniform accompanied by two unidentified people, a man and a woman. From the facial resemblance, one can safely assume that the lady is related to Anthony in some way.

What is interesting about this picture is that the woman has her arm draped casually around her male companion's shoulder. This would have been frowned upon in pre-war times as any public display of affection was rare, and usually people posing for photos looked very serious and formal.

Anthony had been granted approval to return to Australia via Malta at his own expense, utilising his leave from 1915. He had left England, travelling overland to Malta, on the 10th of December 1918. It is unclear how long he stayed in his country of birth. While in Malta, it is possible that he met up with his friend Eligio Castaldi, who arrived in Malta a few weeks later, on the 7th of January, as reported in '*The Daily Malta Chronicle*' of the next day:

*'Mr Eligio Castaldi, PL, second son of the late Professor Salvatore Castaldi LL.D who had settled in Australia six years ago and had held important positions in the Civil Service, both in the Commonwealth and Fiji Islands, arrived here yesterday on leave from the front where Mr Castaldi has been doing work for some time as Corporal in the Australian Forces.'*

Anthony's next destination was Port Said in Egypt, where he was moved to a transit camp. His service record indicates that originally, he was intended to leave Egypt for Australia on the ship *Burma*, on the 26th of July 1919. No reason is given why his journey was postponed. He ended up staying a further four months, sailing on the *Ypiranga*, originally a German ship, that had been ceded to Britain as a war reparation, on the 27th November, arriving in Australia on the 9th of January 1920, after having spent more than four years away from his adopted home. Upon arrival, he travelled by train to the first Military District in Brisbane.

A medical report, dated 17th June 1920, refers to Anthony's hospitalisation for shell shock in France in 1916, and because of a sprained ankle in October 1918. It was confirmed that the disability was due to military service. His condition and progress at the time of the report were assessed as normal. The Medical Officer and Anthony's case manager, Lieutenant Colonel S.H.

Seecombe did not recommend that Anthony be discharged as permanently unfit for general service. This was concurred by the officer in charge of the Hospital Major G.H. Brandis. However, on the same day, the Medical Board, presided by Lieutenant Colonel J. Espie and Major Eustace Russell as member, did recommend that Anthony be discharged. The board also concluded that Anthony's working capacity was declared not to be affected at all by the disability. He was deemed *'fit for work or vocational training'* and that *'No further treatment in Sanatorium, Convalescent home, Asylum or any other institution or surgical appliance'* was required.

The Department of Repatriation was informed of the Board's findings on the 29th of June 1920, after the decision was confirmed by Colonel Alfred Sutton, the Director General of Medical Services. On the 16th August 1920, Anthony was discharged from the A.I.F.

In his demobilisation form, completed in December 1919, Anthony advised that in his last occupation before enlisting, he was paid £240 *per annum* as an assistant surveyor with the Commonwealth Government, and that he had a promise of employment for the same job. He declared he was not a member of a trade union, and required the assistance of the Department of Repatriation to return to work.

In July 1920, Anthony's application to receive £10 worth of tools related to his profession was approved. In his application, Anthony stated that he was single and debt-free. He declared he had no physical disability, was not in receipt of any pension and did not own any property. His employment history included a five and a half-year stint as an assistant surveyor, and one year as a computing draughtsman. The equipment he applied for included set squares, a mathematical set, brushes, a parallel ruler, and a surveyor's chain scale. After five years, the tools would become his if the Department was satisfied that proper use was made of them. At the time, Anthony was living in Brisbane, care of Mrs. Cowie in Westbourne Street. He planned to leave for Darwin to resume his previous occupation with the first available boat. He left Queensland on the 15th of August and started work on the 1st of September. Three months later, in the absence of the warden, Anthony was given that position for one of the gold mining districts in the Northern Territory.

In the meantime, while Anthony was trying to put everything behind him and settle back to a 'normal' existence, Charles's and Waldemar's families were still grieving and coming to terms with the futile and untimely deaths of their loved ones. To make things worse, long after the war ended, they had to battle against the slow and frequently frustrating red tape of the various departments involved in matters such as the recovery of the dead men's personal effects. The endless correspondence to and fro would have prolonged their anguish and sabotaged their yearning for some kind of closure. Every official letter they received would reopen their wounds and remind them of their painful loss all over again.

On the 1st of August in 1919, Barbara wrote to the Kit Depot in Melbourne, asking for information on Waldemar's kit and personal belongings:

*'His mother is Next of Kin, being Mrs Carolina Beck 91 Sda Ridolfo Sliema Malta. The last letter I received from Malta they had not received his kit or had any word from the Military in connection with it. So they asked me to make inquiries at Perth for them, which I did. But the officer give (sic.) your address saying you will have all the information.'*

In a letter dated 14th August, Barbara was informed that:

*'... no personal effects of the late No.6969 Private W.Beck, 16th Battalion, have been received at this office to date.*

*The kit, being part of the military equipment and the property of the Government, would not be returned, but any articles of sentimental value would be extracted and forwarded.*

*In the event of any of the deceased's effects being received by this Branch, they would be promptly transmitted to his mother, in accordance with his testamentary instructions, but it is considered probable that any articles recovered would be forwarded to her direct from AIF Headquarters London.'*

A few months later, a parcel containing *'Effects received from the field'* was sent to Waldemar's mother from the A.I.F Kit Store in London. The sealed package contained two discs, a metal cigarette case, two wooden cigarette holders, a belt, a nail file, scissors, two knives, a wallet, a diary, a

Y.M.C.A wallet, a purse, a gold ring, letters, photos, and cards. His mother also received Waldemar's prayer book, rosary beads, two crucifixes and four religious charms.

Carolina returned the acknowledgment slip on the 3rd of November:

*'I hereby acknowledge receipt of one package No. D/S 41346 containing effects of the late 6969 Beck W 16th Battalion AIF.'*

For Carolina, it would have been of little comfort to receive her deceased son's meagre possessions, but it was all she had left of him. The Becks sent Waldemar's diary to Barbara as a keepsake, which item was preserved with love through the years by the Gugeri family. It was Barbara's niece who kindly passed on a copy, which is reproduced in the previous chapter.

The Bonavia family not only never received any news about the location of Charles's body, but there was also confusion about the whereabouts of his personal belongings. On the 25th of April 1918, the third anniversary of Charles's death, the law firm Lohrmann and Tindal, responsible for the administration of his estate, wrote to the Base Records Office in Melbourne, advising that the deferred pay had already been collected by the Curator of Intestate Estates. The solicitors also advised that the West Australian Trustee Executor and Agency Company Ltd of Perth had applied for administration of Charles's estate in order to obtain payment of his deferred pay, so that it would be passed on to the next of kin. In the same letter, the solicitors also referred to Charles's belongings, which were still missing:

*'We shall be glad to know whether you have any effects belonging to the deceased soldier.'*

In previous correspondence between the two offices dated April 1918, the law firm was sent a certificate of report of death:

*'Report of death:*

*No 157 Private Charles Emanuel Bonavia, 11th Battalion, AIF was killed in action, Gallipoli peninsula, 25th April 1915.*

*Cable no.MFC42028 from the Officer Commanding Intermediate Base, dated Alexandria, 16th May 1916, confirmed by Army Form B 104-52 and Army Form B 2090A.'*

While enclosing the said report, the Records Office asked the solicitors for

Charles's will or probate to be passed to the District paymaster in Perth for notation so that his military affairs would be dealt with according to his testamentary instructions.

Unlike Waldemar, Charles had died intestate. The absence of a will complicated things and prolonged the process further as the Curator of Intestate Estates had to be involved. The State Records Office of Western Australia also possess a copy of a power of attorney signed by Charles's father on the 20th of November 1917, witnessed by the Notary Eduardo Calleja Schembri. Mr. Bonavia appointed a trustee company to collect money and any other effects from the curator, and authorised it to perform other related acts in relation to the deceased's property. The sum total of Charles's funds amounted to £124 18 shillings and 9 pence, comprising £114.10.9 held at the Commonwealth Savings Bank and £10.8 military pay. He had no property and did not owe any money.

Replying to the law firm on the issue of Charles's possessions on the 7th of May 1918, the Base Records Office stated that:

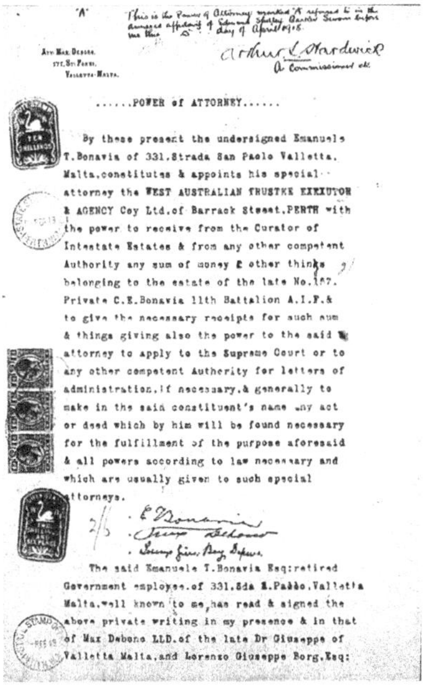
"A"

Arthur L. Hardwick
A Commissioner etc.

......POWER of ATTORNEY......

By these present the undersigned Emanuele T. Bonavia of 331, Strada San Paolo Valletta. Malta, constitutes & appoints his special attorney the WEST AUSTRALIAN TRUSTEE EXECUTOR & AGENCY Coy Ltd. of Barrack Street, PERTH with the power to receive from the Curator of Intestate Estates & from any other competent Authority any sum of money & other things belonging to the estate of the late No.157. Private C.E.Bonavia 11th Battalion A.I.F.& to give the necessary receipts for such sum & things giving also the power to the said attorney to apply to the Supreme Court or to any other competent Authority for letters of administration, if necessary, & generally to make in the said constituent's name any act or deed which by him will be found necessary for the fulfillment of the purpose aforesaid & all powers according to law necessary and which are usually given to such special attorneys.

The said Emanuele T. Bonavia Esq: retired Government employee, of 331. Sda S.Paolo, Valletta Malta, well known to me, has read & signed the above private writing in my presence & in that of Max Debono LLD. of the late Dr Giuseppe of Valletta Malta, and Lorenzo Giuseppe Borg, Esq:

63. *Power of attorney signed by Mr. Bonavia*

*'no personal effects of the late no.157 Private Charles E. Bonavia, 11th Battalion, have been returned to this Office to date but any coming to hand will be promptly forwarded to the Curator of Intestate Estates, Perth WA.'*

To understand this delay, which undoubtedly would have created more anxiety and heartache for the Bonavia family, we have to go back to 1917 when correspondence was exchanged between the A.I.F headquarters in Egypt and London, the Australian Navy, the Department of Defence, and the Base Records Office in Melbourne.

In September of that year, the District Naval Officer in Fremantle had suggested to the 5th Military District to:

*'send the package of personal belongings of Private C.Bonavia in charge of O/C troops on Transport KYARRA leaving Fremantle about September 14th for Egypt with instructions to that Officer to have it left at Malta, or if the ship is not calling there, at Alexandria or Port Said, to be forwarded by the Military Authorities at that port, by first opportunity to o/c Malta for delivery.'*

On the 17th of the same month, the Department of Defence in Melbourne was advised that a number of personal belongings of Charles:

*'have been packed in a strong case bound with hoop iron, and forwarded by the transport Kyarra sailing from Fremantle 17/9/17 fully addressed to Madame Bonavia who has been duly advised and keys of the trunks forwarded per Registered Post.*

*The advice of the District Naval officer was obtained as to best means of forwarding the package to Malta, as it was (after due deliberation) not considered that Defence Memo No.54249 applied to islands in the direct trade route.'*

This case was received at the Australian Headquarters in Egypt, but it stayed there gathering dust for months on end since no instructions as to its disposal had been given. A number of letters went back and forth between the various departments and it seemed that Charles's belongings would get lost among all the bureaucratic red tape. On the 31st of December 1917, the A.I.F headquarters in Cairo wrote to the office of the Base Records in Melbourne, asking for clarifications as to the destination of the said case:

*'Some time ago a case was received here from Australia, ex HT A55 Kyarra, addressed to OC AIF Kit Store London for Madame Bonavia 331 Strada San Paolo Valletta Malta.*

*A Memo was sent to Administrative HQ AIF London asking if they had received any instructions regarding this package and a reply has been received to the effect that no communication has yet been received from you on this matter.*

*Would you therefore please furnish instructions as to the disposal of the case above referred to?'*

Two months later, on the 4th of March 1918, the Base Records Office forwarded copies of communication regarding the ultimate destination of the case. And so the paper trail continued until at last, in June, Charles's family received the much-awaited parcel containing some of his personal belongings with a note that these were *'the property of Private C.E Bonavia, deceased'*. His sister Maria signed and returned the acknowledgement card to the A.I.F Kit Store in London on the 27th of June. The package contained a damaged mirror, a razor strop, photos, cards, and a negative.

However, these were not the items packed in the case stuck somewhere in Egypt as the inventory of its contents listed different objects: eighteen books, personal papers, one watch (out of order), one gold ring (with yellow stone), one 50 ft measuring tape, one small box sundries, one life savings medal (engraved C.E. Bonavia), one hair brush, one clothes brush, one baby's hair brush, twelve photos (four unmounted), sixteen cards (postcards), two leather-bound weights, one travelling trunk, and a University of Malta parchment. Therefore, one can deduce that the package received by the family contained items found in Gallipoli, although the family were not given any information as to how and from where they were retrieved. On the other hand, the items in the case that was shipped from Australia were belongings that Charles had left behind in Perth. No evidence could be found as to whether the case ever reached the family in Malta.

Two years later, on the 17th August 1920, Charles's family was sent the 1914/1915 Star medal *'as one of the mementos of the gallant service rendered by your son the late No.157 Private CE Bonavia, 11th Battalion, with the Australian Imperial Force.'* They were also advised that:

*'The other medals, etc. will be transmitted as soon as they are available, no application being necessary.'*

The 1914/15 bronze Star has a central scroll showing the dates 1914-15, and, like the 1914 Star, awarded to those who had served in France between August and November 1915, has a ribbon of watered red, white, and blue. It was issued to all those who saw service between 5th of August 1914 and 31st of December 1915. Similar to the 1914 Star, it could not be awarded on its own, and the recipient was automatically entitled to the British War and Victory Medals.

Since the medal had been addressed to Charles's father Emanuel as his next of kin, the office of Base Records queried the fact that the receipt form was not signed by him. In a letter dated 24th of May 1921, the office wrote:

*'Will you kindly favour me with advice as to whether Mr E.T Bonavia still survives and also what relationship you bear to the deceased soldier. The instructions governing the disposal of War Medals prescribe that they may only be handed over to blood relations in the following order:*

*"surviving father, mother, eldest surviving brother, eldest surviving sister, eldest surviving half-brother, eldest surviving half-sister etc"*

*unless there are good and sufficient reasons for varying the procedure.'*

The receipt had in fact been signed by Charles's mother Giorgina in October 1920. On the 12th of July 1921, Mrs. Bonavia advised that:

*'Dear Sir,*

*In answer to your letter dated 24th May, I Giorgina Bonavia am the mother of the deceased soldier No.157 C.E Bonavia 11th Battalion. His father has been dead for some years and therefore the reason of my signature.'*

Charles's father had passed away in 1918, without knowing exactly what had happened to his beloved son. Unfortunately, his mother was reminded again of this cruel fate when she was sent the pamphlet '*Where the Australians rest*' in October 1921, even though Charles would never have a formal resting place since his body was never found. In September 1919, she had written to the Imperial War Graves Commission after being requested to provide the wording for a proposed tablet to commemorate Charles. She asked *'to be informed of the exact locality where the tablet will be placed that I might perhaps someday be able to visit the spot which is to be my dear son's grave.'*

Lea Harding née Stilon Depiro, whose mother Evelyn was Charles's sister, recalls visiting her grandmother Giorgina every Sunday. Years after her beloved son's death, Giorgina still hoped that perhaps Charles was missing, not dead, and that one day he would come back. Looking longingly at his photo, she would cry and tell her granddaughter *'Charlie għad jiġi' ('Charlie will come back'.)* Sadly, Giorgina would encounter more tragedy in her life, as she not only lost Charles and her husband just a few years apart, but her daughter Evelyn, Lea's mother, also predeceased her, dying when she was only fifty. Lea, who lives in Sydney, always mentions Charles to her children and grandchildren, and attends the dawn service religiously every year to commemorate the uncle she had never known but whose memory and legacy she holds dear to her heart. A friend of her daughter visited Gallipoli on the 100th anniversary of the landing, and paid tribute to Charles at the very place and a whole century after to the day when his young life had ended so prematurely.

The British War Medal was sent to Mrs. Bonavia on the 20th of October 1921, followed by the Memorial scroll and plaque in February 1922. In September, she received Charles's Victory Medal.

Waldemar's family received his British War Medal in August 1921, while the Memorial plaque and scroll were sent in November 1922, followed by the Victory medal which his mother received in June 1923. He was not eligible for the 1914/15 Star as he had enlisted after 1915.

The silver British War Medal 1914/20 was automatically awarded to those who qualified for the 1914, 1914/15 Stars and Victory medals. It was also issued singly to personnel who did not see any fighting. The ribbon has a wide central orange stripe edged with stripes of white, black and blue, with an image of Saint George on the reverse.

G v R I

Dieu et mon Droit

HE whom this scroll commemorates was numbered among those who, at the call of King and Country, left all that was dear to them, endured hardness, faced danger, and finally passed out of the sight of men by the path of duty and self-sacrifice, giving up their own lives that others might live in freedom.

Let those who come after see to it that his name be not forgotten.

Pte. Waldemar Beck
16 Bn. A.I.F.

***64. Waldemar's memorial scroll***

The ribbon of the bronze Allied Victory Medal 1914/18 is of a watered rainbow design, having a central red stripe flanked by yellow, green, blue and violet stripes in a vertical pattern. Awarded to those who had served between the 5th of August 1914 and 11th of November 1918, it has a winged figure depicting Victory holding a palm branch.

The Memorial Plaque, also known as *'dead man's penny'*, was given to the next of kin of those who had lost their lives on war service both during and for a short while after the war. About 13 cm in diameter, it bears the inscription *'He Died for Freedom and*

*Honour'* and the name of the person commemorated is cast in raised capitals. The plaque was accompanied by a scroll which bears the inscription: *'He whom this scroll commemorates was numbered among those who, at the call of King and Country, left all that was dear to them, endured hardness, faced danger and finally passed out of the sight of men by the path of duty and self-sacrifice giving up their own lives that others may live in freedom. Let those who come after see to it that his name is not forgotten.'*

Below this inscription in ink is recorded the name, rank and unit of the deceased.

The medal presented to the volunteers in Malta showed the regimental badges of the ANZAC, Canadian and South African forces. Volunteers in hospitals in Malta were awarded with a special cross inscribed with the words: *'To commemorate your share in the Great War 1914-18.'*

The Victoria Cross, the highest military decoration, was awarded to 64 Anzacs, 9 at Gallipoli, and 53 were awarded between June 1916 and October 1918 on the Western Front, in France and Belgium. Albert Jacka, the hero of Gallipoli and recipient of the very first VC of the conflict, was reported by the press that *'since the war's end Jacka had been on the Mediterranean island of Malta.'* This was denied by Jacka who was quoted by a reporter saying: *'I was never near the place.'* Jacka had also received a Military Cross in France.

A Maltese Anzac of the 19th Battalion, Private 4695 Paul Debono, was awarded the Distinguished Conduct Medal in June 1918. His citation reads:

*'On the 5th June 1918, at Morlancourt near Albert, the enemy raided our trenches and entered them in places. Pte Debono was then carrying a message overland to the front line. On reaching his trench, he was faced by 6 Germans who showed fight but Debono immediately bayoneted one and his determined attack cowed the others that they at once surrendered to him.'*

Paul survived the war and lived to the ripe old age of eighty until 1967 at Maroubra in NSW.

Anthony received the 1914/15 Star, the British War Medal, and the Victory Medal which he returned in 1922 to be amended as his surname was incorrectly inscribed as *'Xuerel'*. In January 1923, the amended medals were sent to the 1st District Base in Brisbane. In turn, that office sent a note acknowledging receipt of the medals.

There is no further correspondence confirming that the medals were in fact then passed on to Anthony. In fact, the last record to be found in his service file, held at the National Archives in Canberra, is a letter dated 18th of March 1967 from Anthony to *'the Secretary Department of Army'*. The home address given is No.7 Castle Street, Fairfield in Queensland:

*'Dear Sir,*

*I wish to claim my medal and lapel badge for service at Gallipoli. The service details were:*

*Number 1055*

*Pte A. Xuereb*

*25th Battalion AIF*

*Yours sincerely,*

*Signed A.Xuereb'*

From the early 1920's, the families of Australian soldiers who had died were sent a form by the Department of Defence in Melbourne, asking them to provide certain particulars. This information was required to build a reference for historians and also *'so that the name and record of every Australian soldier who fell may be inscribed in the great Honour Roll of the Australian people in the Memorial Museum.'*

*' ............The name of every man who gave his life for Australia will be recorded in the Australian War Museum, which will be the Memorial of the Australian Nation to its honoured dead. The names will stand there for endless generations of our children to read them.'*

Charles's sister Melita completed the form, giving information such as his education and that he had arrived in Australia when he was twenty-four years old. Under the heading *'Any other biographical details likely to be of interest to AIF Historian or of his Regiment'*, Melita remembered her brother fondly:

*'He was very much loved and considered very clever by his regiment.'*

Another of Charles's sisters, Marie, migrated to Sydney with her husband in 1926. When the ship docked at Fremantle, she went to look for Gertie at the address where the family used to write to her. Although Gertie in the mean-time had married, it is evident that Charles still had a special place in her heart, as this letter from Marie to her sister Melita testifies:

*'Oh Melita, she was so delighted to see me, poor girl, and she saw I was very much like Carlo. She is married. She is Mrs Andrews now. I have received a letter from her since I have been here again telling me how pleased she was to see me.'*

The two women would have talked about their memories of Charles and undoubtedly shed a tear together, sharing their grief over a man they had both loved:

*'.....I was very pleased that I could kiss and embrace the woman Charlie had loved so much.'* (Letter dated 4th of April 1926)

The information on Waldemar collected by the Australian War Memorial was provided by his brother George, who also gave Barbara's name as another person to whom reference could be made by historians for further details. George wrote that besides being a land surveyor and architect, Waldemar was also trained as a *'motorman'*. As additional biographical details, George provided the following information:

*'The deceased entered the Malta University in 1906 where in 1910 he obtained the degree of B.E and L.S. & A. In 1912 he left Malta for W. Australia. In 1916, he enlisted and in the same year he went to the front. On the 4th of July 1918 he received a gunshot wound over the left eye. He was sent to the 8th Military Hospital, France where he died.'*

The 1914-1918 Roll of Honour at the Australian War Memorial in Canberra, for which these details were collected, comprises 188 bronze panels on which are displayed the names of over 61,000 Australian servicemen and women who died during World War I. It is significant to point out that no ranks or awards are listed next to the names, indicating that each life sacrificed was no less important than the other. The Roll also brings home the fact that these were all individuals and not numbers. Charles is commemorated in panel 61, while Waldemar's name appears on panel 78. Charles's name also features on panel 33 of the Lone Pine Memorial at Gallipoli, and together with Waldemar's, at Kings Park Memorial in Perth. As part of the Anzac centenary commemorations, the names from the panels, including those of Charles and Waldemar, will be projected one by one for thirty seconds each on the exterior of the Australian War Memorial's Hall of Memory. In this commemoration, running from the 4th of August 2014, the day when war was declared on Germany, until the 11th of November 1918, each name will be displayed about thirty times.

As part of the project *'Gallipoli Dead from Western Australia'*, Charles's name was also displayed on the War Memorial in Perth before the Anzac Day service in 2015, and read out by children from Greenmount Primary School, next to the old Blackboy Hill site. Charles was also represented by one of the 1023 red helium balloons released from the Bell Tower in Perth, commemorating the 1023 men from Western Australia who perished in Gallipoli. In addition, his name was painted on a surfboat that took part in a race at Gallipoli to mark the 100th anniversary. At the launch of the Cottesloe and Trigg Island surfboats that competed in the said race, Shannon Lovelady, one of the brains behind the Gallipoli Dead project, also did a reading about Charles, referring to the Cottesloe Surf Life Saving Club bronze medallion he had been awarded in March 1914. This medal was one of the items packed in the case that had been held up in Egypt.

The Australian War Memorial, considered one of the world's great national monuments, was the brainchild of Charles Bean who had dedicated twenty years to write the Anzac story in *'The Official History of Australia in the War 1914-1918'*. He had thought of the idea of building a monument to perpetuate the memory and sacrifice of the Anzacs after witnessing the horrors on the Western Front in 1916:

*'Many a man lying out there at Pozières or in the low scrub at Gallipoli, with his poor tired senses barely working through the fever of his brain, has thought in his last moments: "Well - well - it's over; but in Australia they will be proud of this."'*

John Treloar, a Gallipoli campaign veteran, worked tirelessly with Bean to establish the Memorial. He was appointed as its Director in 1920, a position he held until his death in 1952. The building of the Memorial was stalled for a while due to the Depression. However, a long-running exhibition of war relics in Melbourne and Sydney had been acclaimed by the media and was enthusiastically received by large crowds. Construction on the Memorial began in 1934, when the Duke of Gloucester planted a pine tree in the grounds from the seeds of a cone sent by an Australian soldier to his mother, which cone he had taken from a Turkish trench at Lone Pine, where his brother had lost his life.

Inaugurated in 1941, besides the Roll of Honour, the Memorial hosts a vast collection of relics and records, photographs, art and archival film that tell the story of the nation's involvement in wars, regional conflicts and international peacekeeping missions. It also houses the Tomb of the Unknown Australian Soldier killed on the battlefields of the Western Front.

Numerous other memorials in honour of the fallen dot the various townships around Australia, ranging from imposing structures in the big cities, like the Shrine of Remembrance in Melbourne, to smaller, more humble monuments in quiet, sleepy towns. At the Shrine in Melbourne, each year on the 11th of November at exactly 11:00 am, a ray of sunlight hits the stone of remembrance highlighting the word *'love'* on the inscription etched in gold *'Greater love hath no man'.*

French and Belgian towns and villages are also dotted with many reminders of the sacrifice of the Anzacs. In Villers Brettoneux, the school was rebuilt with monies raised in Victoria. A street is also named *'Rue de Melbourne'* as it was Victorian battalions that rescued the town from German hands between the 24th and 25th April 1918, on the third anniversary of the Gallipoli landing. The Australian Park Memorial in Hamel, where Waldemar lost his life, commemorates the engagement of more than 100,000 Australians who served in France.

In its role as *'Nurse of the Mediterranean'*, Malta became a resting place for 308 Anzacs: 229 Australians and 79 New Zealanders. There are also 1,542 British and a number of French, Indian, and Egyptians soldiers, as well as 26 Turkish prisoners buried there. Over the years, many would visit Malta to pay respect to their fallen relatives. *'The Daily Malta Chronicle'* of the 14th of June 1916 reported that:

*'Relatives and friends of our heroes at rest in Malta may be assured that the graves of their loved lost are well looked after and will not be allowed to lack*

*flowers in token of loving memory from over the seas.'*

Two stained-glass windows, depicting Christ the Light and Christ the Good Shepherd, were donated by Australian and New Zealand Methodists convalescing in Malta to the Methodist Church in Floriana in 1921, to commemorate their mates who had their resting place on the island. When the Church was closed in 1974, these were moved to St. Andrew's Scots Church in Valletta. In April 2015, the Maltese RSL sub-branch of South Australia unveiled a memorial with the names of South Australian soldiers who are buried in Malta.

While the graves of the foreign soldiers were tended with great love and care, Elio Lombardi, author of the book *'L-Ewwel Gwerra Dinjija' ('The First World War')*, complained that after three-and-a-half years since the end of the war, there was still no monument to commemorate the 1,000 Maltese who had perished in the conflict. While asking readers for details of their dead relatives to publish in a future edition, he criticised the authorities:

*'...sal-lum għaddeu quasi tliet snin u nofs li spiċċiat il Guerra u għada ma dehret l'ebda ġebla imkar imnakxa bil mus, sabiex tibka memoria ta dauc li ħalleu ħajjithom għal curuna, mita beħsiebhom jagħmlulhom dan il monument? Jakau la jiġi secolu iħor?'*

The author goes on lamenting that future generations will have no clue about their compatriots who fought and gave their life for them.

The war memorial at Floriana once recorded the names of a number of Maltese who had died in the war, and also included the Maltese who had died while serving with the Australian forces. Unfortunately, when the road was widened, their names were removed.

A monument to commemorate the strong ties between Malta and the Anzacs was inaugurated on the 25th of May 2013 at the Argotti botanical gardens in Floriana. This was the result of the joint effort of the Maltese-Australian Association of Malta and the Friends of Australia, with funds raised in both countries, to underline the long-standing links between Australia and Malta.

In November 2015, a plaque was unveiled by the Maltese Ex-Servicemen Association NSW Sub-Branch at Pendle Hill in NSW, containing the names of fifty-three Maltese Anzacs, including Charles, Waldemar, Anthony and Eligio and also five who were identified by researcher Mark Caruana to have enlisted in New Zealand. Among the descendants and family members present, there were Charles's niece, Lea Harding, and Peter Castaldi, who commemorated both his father and his father's friend, Anthony or Tony, who he still remembers with great affection.

On the first anniversary of the Gallipoli landing in 1916, a commemoration presided by the Governor Lord Metheun was held at the Pieta` Military Cemetery. A memorial service, attended by the Archbishop of Malta, Mauro Caruana, was also held at the Addolorata cemetery for the Catholic soldiers. Both services were attended by nurses serving in the military hospitals and convalescing soldiers to salute their fallen mates.

Since then, both Anzac and Remembrance Day (which before the Second World War was known as Armistice Day) have been celebrated every year in Malta, with a wreath-laying ceremony at the Pieta` military cemetery and the War Memorial in Floriana respectively. These are attended by Malta's highest officials and diplomatic representatives from Australia, the United Kingdom, New Zealand and France. Commemorations were even held during the period when the island was under heavy air bombardments during the Second World War.

Sir Hubert Opperman, the famous cycling champion and former Australian High Commissioner to Malta, in one of his Anzac Day speeches said:

*'While I speak to you of our Anzac Day it is with a warm feeling of using the word as an all-covering canopy over both our countries. Maltese were alongside the Australians at Gallipoli, and in equal danger, for no one was immune or better sheltered than anyone else, and as our casualties mounted astronomically, Malta tended our wounded, gave hospitality to the convalescent, embraced our death in her earth...'*

In 1930, Malta also served as the location of the movie released in the United States as *'The Battle of Gallipoli'*, based on the novel *'Tell England'*, published in 1922 by Ernest Raymond, who had served as a chaplain at Gallipoli. One of the directors, Anthony Asquith, was the son of the British Prime Minister. The Għajn Tuffieħa and Mġiebah beaches provided a backdrop for the landings. Extras were provided by the Navy and Army, while the sea vessels were supplied by the Mediterranean fleet.

Over the years, various reunions of Gallipoli veterans occurred in Malta. One such reunion happened in 1936, when around 600 veterans converged on the island, led by Field Marshal Sir William Birdwood, who acknowledged that the Australians and New Zealanders still remembered the hospitality and excellent service they had encountered on the island.

In April 1990, seventy-five years after the landing, a historic journey took around fifty veterans of the Gallipoli campaign from Australia all the way back to the Peninsula for a commemorative ceremony. There, they met with Turkish veterans who welcomed them as friends in a congenial atmosphere, a far cry from the fierce fighting as enemies that had transpired seventy-five years earlier. They attended a very moving Anzac Day dawn service at Lone Pine, which marked the last ceremony to be attended in the Dardanelles by survivors of the original landing. One of the veterans also celebrated his 103rd birthday there.

In April 2006, the documentary *'Australian Story'* on the ABC channel featured the Mackay North State High School's annual Spirit of Anzac Commemorative tour of Gallipoli and other World War I and II battlefields, memorials, and cemeteries. Later that year, the tour would be taking the students to Gallipoli, Flanders, Somme, Pozières, Bullecourt, Paris, and Singapore. After every trip, the young participants all came back as different people, vowing that they would live the lives the fallen young men had missed out on.

Impressed with the enthusiasm of the students and the commitment of the teachers to bring history to life, I contacted one of the organisers, Mike Goodwin, asking him to commemorate Charles in Gallipoli and possibly to visit Waldemar's grave at the Boisguillaume Communal Cemetery. Mike wrote back saying that, although regrettably it was impossible to include a visit to Waldemar's grave due to time constraints, he was more than happy to have Charles commemorated at Lone Pine Memorial at a special service. Interestingly, Mike told me that he also had a Maltese connection as two uncles of his brother-in-law's grandmother are buried in Malta. He had contacted John Mizzi in 2000 who photographed their graves and also ran an extensive article on the school's project in a special Anzac Day feature in the Times of Malta. In his turn, Mike also graciously managed to obtain a photo of Waldmar's grave for me through one of his contacts Matt Smith, the Director of the Australian War Graves Photographic Archive.

The Maltese connection did not stop there though. Mike put me in touch with Jack Formosa, a student of Maltese background. When I sent Jack a photo of Charles so he could put a face to the name, he wrote:

*'I now have an idea of how he looks. These photos put in perspective how young this man was and how brave he must have been.'*

Jack was very happy to commemorate Charles in Gallipoli on the 26th of September, where he laid a poppy in his honour and read a citation about his life. He had also laid a poppy on the Roll of Honour panel near Charles's name at the Australian War Memorial on a school trip to Canberra the previous July. In France, Jack also commemorated a relative of his who had won the Victoria Cross. In Jack's citation about Charles, he mentioned Charles's move to Australia, his enlistment with the 11th Battalion, and his tragic death:

*'I am commemorating Charles on behalf of Mrs. Gioconda Schembri of Melbourne and his family who still live in Malta. They will never know exactly what happened to Charles but he will be in their hearts forever more.'*

Below is a very touching note Jack wrote to me after returning from his trip:

*'I must thank you once again for giving me the honour to commemorate Charles Bonavia at the Lone Pine Memorial. Standing in this special place, and reading my eulogy was a great honour. I believe that Charles's presence was with me while I stood there. The most undescribable feelings came over me at this time, feelings that I will never forget. Charles has become a part of my life in the last 12 months and he will be forever more.'*

In another email, he describes how emotional it was to find Charles's name on the memorial. I had gone through the same overwhelming experience when I visited Gallipoli six years earlier.

*'...I cannot begin to tell you the feelings I had when I found Charles's name on the Lone Pine Memorial. This name on the wall was so powerful and I was honoured to be standing in front of his name while reading his eulogy.'*

In a letter Mike sent me after the tour, he reiterated what an unforgettable experience it had been for the students:

*'Jack delivered a lovely commemoration during our special service and the*

*65. Jack commemorating Charles at Gallipoli*

*group was very proud to honour Charles this way...it was a great honour for us to visit and commemorate Charles and I am sure he enjoyed the company of these 28 young Australians for that short time that they brought a piece of Australia to him.'*

Indeed, the students and teachers offered a fitting tribute to Charles at the very location where ninety-one years earlier, he had died alone and without his body ever been found to be given a dignified burial. At last, he was given the proper commemoration he deserved.

Charles continued to impact Jack's life long after the trip. On the 5th of February 2007, he emailed me the following:

*'I also got a frame which holds my four best mates in it looking over my bedroom, Charles being one. I still pick up my album and shed the odd tear or get butterflies when looking over the memories. I made a promise to my boys that I will come and visit once again in the future and I never break my promises.'*

Years later, when Jack moved to New Zealand, he took Charles's framed photo with him.

In the post-war years, Anthony, the last survivor of the trio of friends, was very active in the Darwin community and social scene. In September 1924, he decorated the hall for a masquerade ball organised by the Darwin Soldiers League, which was deemed a huge success. '*The Daily Standard*' of the 29th December 1924 mentioned that he supervised returned soldiers in the decoration of the soldiers' memorial hall in preparation for a wedding. Anthony was granted long service leave towards the end of 1925, returning to work in October 1926.

'*The Townsville Daily Bulletin*', in its July 1936 column '*Darwin Notes*', reported that Anthony, as the longest-serving member of staff at the Lands Office, presented Miss Eileen Granger with '*a beautiful crystal water set of 14 pieces*' on behalf of the employees as a farewell gift in view of her approaching marriage. The same column, in October 1931, had reported that:

'*There is again an epidemic of sickness about the town of Darwin, contingent with the change of season. Messrs R.H Bowman, dentist and 'Tony' Xuereb, draughtsman of the Lands Department were recently in hospital. Many aboriginals are also afflicted including some natives in Government employ.*'

He was also heavily involved with the Anzac Sports Day held on Anzac Day in 1928, where we find him in charge of the sweets stall. In 1934, we see him involved with the committee of the Aero aviation club. Although Anthony by now had been in Australia for over twenty years, except for the four years he spent serving overseas, he had not forgotten his Maltese roots and still kept in touch with the Maltese community. In March 1935, he attended a reception organised by the Maltese Farmers and Settlers' Association in honour of Robert Hamilton and his wife who had migrated from Malta. Hamilton was the former Minister for Labour and had been a member of Gerald Strickland's political party.

Anthony still kept in touch with his good friend Eligio Castaldi and visited him often in Sydney. Eligio's son, Peter, recalls these visits with nostalgia:

'*I remember him well from many visits to our home when I was a boy. He came every year during his annual leave from his job in Darwin, arrived and left by ship - quite fun for a young person as we saw him off, streamers and all!'* (Email dated 12th of May 2015).

*'They used to have what seemed like tremendous arguments after dinner and some wine but I know it was a small boy's reaction and their warm friendship was evident.'* (Email dated 6th of November 2009).

The now defunct Maltese newspaper published in Australia, *'Malta News'*, reported on the 1st of May 1945 that Eligio, who at the time was a solicitor with the Sydney firm Messrs. Hickey & Co., had travelled to Melbourne the previous month, accompanied by Anthony. The newspaper describes them as having been *'comrades in the first AIF'*. During their visit, they had a meeting with Mr. Emanuel Gatt, President of the Maltese Social Society, and other committee members. This not only proves their continued interest and contribution to the Maltese community, but is also a testimony to their long and strong friendship.

In July 1937, Anthony was one of around forty civil servants who volunteered to work on the wharf unloading food, following a strike by the dockworkers. According to a newspaper report of the time, the volunteers toiled from day till late at night in the boiling sun:

***66. Peter Castaldi and Charles N. Mifsud, President of the Maltese RSL Sub Branch (NSW), next to the plaque commemorating the Maltese Anzacs***

*'With sore hands and backs, volunteers worked until a late hour last night unloading the Marella. Transplanted suddenly from the quiet of office duties to the grind of a wharf labourer's job, many of the volunteers suffered severely yesterday. In addition to the loading and unloading of slings, the men had to push the railway trucks to and from the sorting sheds when the locomotive drivers joined the strikers. One man collapsed and the majority were exhausted when they left the jetty late at night'* (*'The West Australian'*, 22nd of July 1937)

At the time, Anthony was almost 49 years old, so the hard labour must have been very taxing for him. Unionists of course were not happy about these 'strike breakers' and the union secretary published the names of the volunteers with sarcastic remarks about each one. When he listed Anthony's name, he remarked:

*'Tony Xuereb, speaks a little English but acts differently.' ('The Northern Standard'*, 27th of July 1937).

By contrast, the same newspaper, in an article published four days earlier, praised the efforts of the *'40 valiant civil servants'* who came to the aid of their community by bringing them much needed food supplies that would have otherwise been stuck on the ship.

In the oral history interview with Deric Thompson, one of Anthony's colleagues at the Lands Department, he remembers that a strong friendship developed between the two of them. He describes Anthony as a traditionalist, *'a great upholder'*. As senior civil officer, Anthony was vice president of the mess. At the time, a number of the staff were *'a bit anti-regimentation'* and once, a mutiny almost occurred as they resisted being taught how to be proper Navy officers. Together with another disciplinarian, the senior staff officer Lieutenant Commander Fowler, Anthony must have had his hands full trying to instill Navy traditions. He managed to persuade the senior naval officer to introduce white trousers in summer like the English Navy. He was also very careful to use proper language such as calling the bedrooms *'cabins'*, in keeping with navy parlance. Furthermore, he was a stickler for rules, such as standing still when the flag was lowered. Thompson recalls Anthony with great respect:

*'........he was a great favourite amongst the Navy, particularly the senior*

*men......because he knew so many of the old Naval people who had been through Darwin.'*

*'......Tony was a great chap. He eventually left the Territory during the war years.'*

Arthur Miller, who became Chief Surveyor of the Northern Territory in 1940, also echoes Thompson's sentiments:

*'.........that was Tony, lovely man.'*

In the early 1940's, Miller appointed Anthony as public relations officer on one of the ships to Sydney, evacuating women and children to safety from the threat of a Japanese attack between 1941 and 1942. He was flown back and then entrusted with evacuating another 200 people.

Anthony must have felt great pride that once again he was doing his duty to help his fellow citizens in yet another war. At the same time, perhaps he would have also thought: *'Wasn't the other war supposed to be the war to end all wars?! So why is it happening all over again? Did Charles and Waldemar and millions like them die for nothing? Was our sacrifice all in vain? When will we ever learn?!'*

Miller moved the Lands and Survey office to Alice Springs where Anthony spent some time *'until about the end of 1944, when he retired with ill health.'* Anthony was *'......starting to fail in health because he enlisted in Darwin for the first great war, went through the war and came back with shell shock, and took up his duties in the old job.'*

Deric Thompson also refers to Anthony's gradual physical deterioration:

*'He just broke down in health in the end. He'd spent so much time in Darwin and he gave out in Brisbane, just after he had won the lottery.'*

Here, Thompson must have confused Anthony with his friend Eligio, as it was the latter who had won the state lottery in August 1935. The Australian electoral rolls between the 1920's and early 40's show Anthony living in Smith Street in Darwin, at least until 1943. So Anthony must have moved to Queensland sometime afterwards. One of the reasons he moved to Brisbane may have been to be close to his younger brother, Charles Alexander (Alex), and his sister-in-law, Jean Ramsay Smith. *'The Telegraph'* of the 23rd of November 1932 reported that Jean was entertained by the office staff of

MacRobertson's Ltd at a theatre party and supper to celebrate her approaching marriage to Alex. They had become engaged in July 1932.

In November 1939, Anthony's brother enlisted with the Australian forces and served for some time in the Middle East. In July 1960, Alex wrote to the Army Medals Section asking for his Second World War medals to be replaced as he had lost them during the Anzac Day march. He also asked whether his brother was entitled to any medals as *'he served as the welfare officer in charge of the evacuation of women and children from Darwin per H.M.A.T. 'Zealandia' which left Darwin 12.30 pm on 20 December 1941.'* The Medals Section replied that *'only those members who were commissioned or enlisted in the Armed Forces or served as a full-time uniformed representative of a philanthropic organisation, such as the Red Cross, are eligible to receive the Campaign Awards.'*

After reading this letter and another that Anthony presumably wrote in 1967 asking for his WWI medals, I was very surprised to discover that he had actually died on the 18th of January 1948! A possible explanation could be that perhaps it was again Alex who wrote the letter after his death, as he was not able to find his war medals among his late brother's belongings.

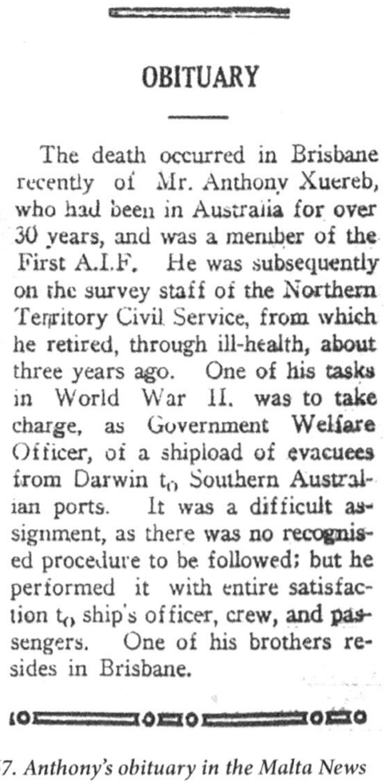

OBITUARY

The death occurred in Brisbane recently of Mr. Anthony Xuereb, who had been in Australia for over 30 years, and was a member of the First A.I.F. He was subsequently on the survey staff of the Northern Territory Civil Service, from which he retired, through ill-health, about three years ago. One of his tasks in World War II. was to take charge, as Government Welfare Officer, of a shipload of evacuees from Darwin to Southern Australian ports. It was a difficult assignment, as there was no recognised procedure to be followed; but he performed it with entire satisfaction to ship's officer, crew, and passengers. One of his brothers resides in Brisbane.

*67. Anthony's obituary in the Malta News*

Anthony's death certificate, issued in the district of Brisbane, shows that he died at Saint Helen's Private Hospital in South Brisbane. He was fifty-nine years old and had retired as a federal public servant. His brother Alex had registered his death, which was caused by a cerebral embolism and a lung abscess. He had never married. His funeral was held two days later at 11:30 am when the procession moved from the Alex Gow funeral chapel at Petrie Bight to the Brisbane General cemetery, also known as the Toowong cemetery, where he was buried in grave number 7A-226B-9. A notice was published in '*The Courier-Mail*' in Brisbane, where Alex and his wife, living at Castle Street in Fairfield, invited their relatives and friends to attend the funeral '*of their beloved brother and brother-in-law.*' Three days later, the same newspaper published the following notice:

'*Mr and Mrs Alex Xuereb wish to thank all kind friends for their expressions of sympathy in their recent sad bereavement, especially the staff of St Helen's hospital for their wonderful attention. Accept this please as our personal thanks.*'

In February, a statutory notice was published by the Brisbane Public Curator in '*The Courier-Mail*' addressed to any creditors who may have had a claim against Anthony's estate.

Anthony's death was also mentioned by '*The Malta News*', which published an obituary in its edition of the 1st of May. The newspaper referred to his stint with the A.I.F and his long career with the Northern Territory Civil Service, '*from which he retired, through ill-health, about three years ago.*' It also mentioned his efforts in the Second World War to take care of a ship of evacuees from Darwin to Southern ports. '*It was a difficult assignment, as there was no recognised procedure to be followed; but he performed it with entire satisfaction to ship's officer, crew and passengers.*'

***68. Eligio in 1960***

Eligio died in 1967, almost twenty years after his friend Anthony.

And what had become of Barbara? It seems that long after Waldemar's demise, the Beck and Gugeri family still kept in touch, their grief for their beloved bonding them forever. Waldemar's great-nephew recalls playing with a set of Chinese dolls which Waldemar's fiancée, Barbara, had sent his second cousin Jo from

Australia. This contact between the two families was renewed decades later, after Mariella Camilleri-Beck and Barbara's niece Barbara started to email each other, bringing the two families together once again.

Barbara lived with her two sisters, Dorothy and Deborah, who were both nurses. She ran the Lattice Tea Rooms in St George's Terrace in Perth. In *'The West Australian'* of the 25th of August 1950, she was featured in an article entitled *'Hints for arrangement of spring flowers'* where she was described as *'well-known for her floral decoration of the Lattice.'*

In an email dated 29th of April 2015, her niece Barbara reminisced:

PERTH FLORAL DECORATION

Miss B. Gugeri arranging a bowl of daffodils at the Lattice yesterday.

*69. Barbara in 1950*

*'Yes that is aunt Barbara - she always looked so elegant and beautiful - I have wonderful memories of going to the Lattice Tea Shop where she and Aunt Dorothy had beautiful flowers and always the most delicious cakes!'*

Barbara died on the 24th of May 1974 in St Vincent's Hospital in Guildford and is buried in East Fremantle. She had never married. Her niece, Yvonne Kelly remembers the day clearly:

*'It was like summer, hot. I think it was a Friday and I said to Dick: 'I think I'll run out and see Aunt Barbara. Do you want to come for a run?' And he said yes, he'd come because he could not drive. And so we went to St Vincent's and I just stayed there for a quarter of an hour and then she breathed her last. And the nun was so sweet, she said to me: 'You have seen the passing of a saint.' She really was a saint'.*

Father Martin Cawley agrees:

*'I personally remember Barbara also, who never married after her fiancé was killed in the First World War, and who gained the reputation of being a real saint.'*

Barbara's niece still feels sad that her aunt had lost Waldemar, who most probably was the love of her life:

*'......All so sad. I think that she never did get over him. He must have been her one great love and no one else measured up. She was a very beautiful person in personality and looks.'* (Email dated 5th of September 2008)

*'So very sad! ...........It does make you wonder how life would have gone under different circumstances. But – such is life and of course death.'* (Email dated 17th of September 2015)

Mariella, Waldemar's great grandniece, shares the same feelings:

*'I'm sure that Barbara and Waldemar are together now at last, and that they are in a better place than this. I do believe that they are watching over us together with Josie and Mary my grandparents.'* (Email from Mariella, 19th of October 2015)

Mariella's grandfather and Waldemar's nephew, Josie, passed away four years after Barbara, on the 7th of August 1978. His parents, George and Vira, had died in 1948 and 1966, respectively. Towards the end of her life, Waldemar's mother, Carolina, lost her ability to speak and on her deathbed, she failed to make herself understood to Vira that she wanted to be handed a silver cross.

70. *Carolina*

Vira always regretted the fact that she had been unable to fulfil her mother-in-law's final wish. Carolina survived her beloved younger son by eight years, dying in 1926 when she was eighty-four years old.

And thus, the curtain falls on the life story of Charles, Waldemar and Anthony, whose promising future and lives were caught up with events beyond their control. Like millions of others from their generation on both sides of the fence, they were the proverbial lions led by donkeys, the victims of a badly-planned war concocted by political and military henchmen in their safe and comfortable boardrooms.

These three remarkable men are not mere names inscribed on cold marble. Their lives mattered and they will be permanently etched in the hearts of their families and loved ones. Praying that I have done justice in honouring them, I know that they will always be a part of my life, and I hope that now they will also hold a special place in your hearts forever.

# The three friends finally at peace

*71. Charles commemorated on the Lone Pine Memorial in Gallipoli*

*72. Plaque on Anthony's grave at Toowong cemetery in Brisbane*

*73. Waldemar's grave at Boisguillaume cemetery in France*

# *With Gratitude*

This book could never have seen the light of day without the invaluable contribution of the families and friends of Charles, Waldemar and Anthony. Lea Harding, Alfred Pisani, Noel Zarb Cousin, Waldemar Beck, Mariella Camilleri Beck, Barbara Barter, Debbie Gugeri, and Peter Castaldi, you were all extremely gracious and generous with your time, patiently answering my never-ending questions. I was privileged to be given access to precious and priceless memories, mementos and photos of your loved ones. Your love and admiration for Charles, Waldemar and Anthony always shone through and I am sure that in turn, our three gentlemen would also be proud of you!

I would like to make a special mention to the late John Mizzi whose book *'Gallipoli: the Malta connection'* was the initial spark that ignited my interest in the Maltese Anzacs. Rest in peace John.

I count myself lucky to call Natalie G. Owens of Divas at Work, my editor and friend *extraordinaire.* A remarkable author herself, Natalie went beyond the call of duty and polished and trimmed my work with a fine tooth comb.

Special thanks to Chris Mahoney who did a fantastic job with the design of the book and cover. He has been ever so patient, gracious and professional in the process of interpreting and transforming my ideas and vision into reality.

Heartfelt gratitude to researcher Mark Caruana for writing the foreword, for his patience and thoroughness in proof reading the various drafts and for his wise suggestions and advice, ever since the idea to write this book was birthed all those years ago.

Thank you to my niece Francesca Cassar who left no stone unturned in meticulously checking all the historical facts and figures.

I am also extremely grateful to my fellow researchers and writers Christine Collins and Lou Drofenik for sharing advice, information and ideas.

The staff at the various archives and libraries in Australia and Malta are to be commended for their professionalism and the great service they provide.

I am indebted to all my family and friends for their continued encouragement and support.

Finally, my gratitude goes to my husband Darren for his everlasting love and patience, and to our son Thomas for being my constant inspiration.

# Bibliography

## ARCHIVES:

**Australia**

Australian Electoral Rolls 1903-1950

Australian War Memorial Missing in Action, Australian Red Cross Society Wounded and Missing Enquiry Bureau files 1914-18 War 1DRL/0428

Battye Library Perth

National Archives, Canberra:

Intelligence report: Waldemar Beck PP14/1 4/7/347

Medical case file: Anthony Xuereb 526,M40735

Passenger lists: C.Bonavia K269, 12 SEP 1911 CHINA, W.Beck and A.Xuereb K269, 10 FEB 1912 MARMORA

Service Records: Charles Bonavia B2455, Bonavia CE, Waldemar Beck B2455, Beck W, Anthony Xuereb B2455, Xuereb A

National Library of Australia (online at www.trove.nla.gov.au):

Digitised newspapers: The Courier-Mail (Brisbane), The Daily News (Perth),The Daily Standard, The Northern Standard, The Northern Territory Times & Gazette, The Swan Express, The Townsville Daily Bulletin, The Weekly Times (Victoria),The West Australian, The Western Mail

Northern Territory Archives Service, Darwin:

NTRS 226, Typed transcripts of oral history interviews with 'TS' prefix, 1979-ct:

Deric THOMPSON, TS 426

Arthur MILLER, TS 285

State Library of Western Australia

State Records Office, Western Australia:

Administration of Charles Bonavia's estate 1918/119

Waldemar Beck service card for employment with Western Australian Government Railway AU WA S23-cons 3393

**Malta**
National Archives of Malta
National Library of Malta
University of Malta:
Melitensia Collection, Roll of Students: 1909-1910, 1910-1911, 1908-1909, 1906-1907, 1904-1905, 1902-1903, 1903-1904, 1900-1901

## BOOKS & RESEARCH PAPERS


Adams Simon, *World War I*, Dorling Kindersley Eyewitness Guides, 2001
Attard Lawrence E., *Early Maltese Migration 1900-1914*, Gulf Press, 1983
*L-Emigrazzjoni Maltija fis-Seklu 19 u 20*, Kullana Kulturali PIN, 1999
Barber Nicola, *Living through World War I*, Raintree Publishers, 2012
Belford W.C, *Legs Eleven*, Perth 1940
Calleja Fr Joe, *L-Imsida Il-Bieraħ u Llum Vol.I*, Veritas Press, 1989
Calleja Paul, *Maltese of the Western Third*, Kallaya Publications, 1993
Carthew Noel, *Voices from the Trenches – Letters to home*, New Holland Publishers, 2002
Caruana Albert & Zammit William, *The Maltese Labour Corps during World War I*, University of Malta, 2011
Caruana Mark, *Malta Passport Applications: Destination: Australia*, 1997
Challinor Deborah, *White feathers*, Harper Collins, 2003
Clarke David, *The Angle of Mons*, Phantom Soldiers and Ghostly Guardians, John Wiley & Sons Ltd, 2004.
Collins Christine, *From Darwin to Dardanelles*, 2015
Ellam-Innes Collection Publication, *Private David John Simcock, 11th Battalion Australian Imperial Force, The bloke with the Pink Top*
Faulks Sebastian, *Birdsong*, Hutchinson, 1993
Gage Nicholas, *Greek fire*, Pan Books, 2001
Gammage Bill, *The Broken Years: Nationhood, Brotherhood and Sacrifice*, Melbourne University Publishing, 2010

Ganado Herbert, *Rajt Malta Tinbidel Vol I*, Klabb Kotba Maltin, 2009
James Robert Rhodes, *Gallipoli*, Pimlico, 1999
Kennedy Jacqui, *World War I letters of Bert Smythe*, www.smythe.id.au
Kerr Greg, *Lost Anzacs – the story of two brothers*, Oxford University Press, 1997
Lombardi Emilio, *Il-Kbira Gwerra Dinjija*, Proprieta' Letteraria u Artistika ta' Adolfo Magro, 1922
Longmore C., *The Old 16th*, Perth, 1921
Macklin Robert, *Jacka VC – Australian Hero*, Allen & Unwin, 2006
Massie Robert K., *Nicholas and Alexandra*, Pan Books, 1967
McDonald Roger, *1915*, University of Queensland Press, 1989
Mizzi John, *Gallipoli: the Malta Connection*, Tecnografica Publications, 1991
Morrow Edgar, *Iron in the Fire*, Hesperion Press, 2006
Rappaport Helen, *Four Sisters, the lost lives of the Romanov Grand Duchesses*, Panmacmillan, 2014
Reid Richard (assisted by Courtney Page and Robert Pounds), *Beaucoup Australiens ici, The Australian Corps in France, 1918*, Department of Veterans' Affairs, 1999
Remarque Erich Maria, *All quiet on the Western Front*, Ballantine, 1987
Sailah Steve, *Stories from Gallipoli*, ABC audio, 2004
Salisbury Harrison, *Black Night, White Snow: Russia's revolutions 1905-1917*, New York, Doubleday, 1978
Scicluna Frank, *Malta and the Anzacs, the Nurse of the Mediterranean*, 2015
Smith Harrison & Koster Adrianus, *Lord Strickland – servant of the crown*, Progress Press, Valletta, 1984
Stone Norman, *World War One – a short history*, Isis Audio books, 2008
Van der Kiste John, *Queen Victoria's children*, ISIS, 1986
Weintraub Stanley, *Silent Night: the Story of the World War I Christmas Truce*, Simon & Schuster, 2014
York Barry, *Empire and Race: The Maltese in Australia, 1881-1949*, NSW University Press, 1990

## NEWSPAPER, MAGAZINE & NEWSLETTER ARTICLES


**Australia**

The Age:

Tippet Gary, *Devotion and loss are written large in a new guide to Victoria's wartime heritage*, 8 November 2009

The Herald Sun:

*The Gallipoli Diaries: the untold stories*, 20 April 2005

Kairos Catholic Journal:

Fahy Robyn, *Catholic women in peace and war*, Volume 21, No.6, 18 April-1 May 2010

Johnstone Tom, *Chaplains at the Landing*, Volume 21, No.6, 18 April-1 May 2010

Hume Moreland Leader:

Anderson Stuart, *From Broady to Gallipoli*, 19 April 2005

Malta News:

Article about Anthony's and Eligio's visit to Melbourne, 1 May 1945, page 2

Obituary, 1 May 1948, page 3

The Maltese Bulletin:

Plowman David, *Malta and the Anzacs*, Maltese Association of Western Australia, June-July 2011

Maltese Cultural Festival booklet:

York Barry, *The Maltese in Australia*, Maltese Community Council of Victoria, 2001

The Maltese Herald:

Bonello de Torella Manuel, *What a coincidence*, 20 March 1990

Baxter Carmen & Caruana Mark, *The Anzac series: A tribute to the fallen Maltese-Australian Anzacs killed in action*, 2010

Bartolo Evarist, *Meta tal-Frawla fetaħ il-bieb tas-safar fl-1883*, 5 October 2010

Our Daily Bread Ministeries:

Cook David, *Freedom is never won without cost: commemorating the ANZAC Day Centenary*, 2015

Wartime: The Australian Experience of War, Official Magazine of the

Australian War Memorial:
Billett Bill, *Malta's AIF connection*, Winter 2001 Issue 14
The Weekend Australian:
Boffa George, *Anzac debts to Malta's brave*, April 1982

**Malta**

The Daily Malta Chronicle
Tal-Ġilju feast commemorative booklet:
Farrugia Charles J., *Infakkru l-memorja ta' Andrew 'Andrea' Camilleri, eroj Imqabbi tal-Ewwel Gwerra Dinjija*, Tal-Ġilju Band Club, 2015
Il-Ħadd magazine:
Mallia Ġużepp OFM Cap, *Anke jien irnexxieli nsib il-qabar tan-nannu tiegħi*, 5 November 2006
The National Archives Newsletter:
*Old Roots New Horizons: the History of Maltese migration, exhibition catalogue* compiled by Charles Farrugia, Friends of the National Archives of Malta, November 2001 Number 3
The Sunday Times:
Muscat Jimmy, *Mellieħa and the Gallipoli campaign*, 9 November 2008
The Times:
Saillard Wayne, *High Commissioner's Malta connection*, 4 May 2010
It-Torċa:
Scicluna Joe, *L-isptarijiet militari u navali f'Malta*, 12 October 2008
Azzopardi Dione, *Il-Gwerra l-Kbira: taqbid qalil fit-trunċieri*, 12 October 2008
*Il-Gwerra l-Kbira: Gallipoli u l-poeti*, 26 October 2008
*Il-Gwerra l-Kbira: il-wasla tat-tank*, 2 November 2008
*Il-Gwerra l-Kbira: il-pajjiżi nvoluti*, 9 November 2008
*Il-Gwerra l-Kbira – ġlied fl-ajru*, 16 November 2008
*Wilfred Owen (1893-1918) il-poeta tal-gwerra*, 16 November 2008
*Il-battalja tas-Somme tagħlaq 93 sena*, 28 June 2009
*Mitt sena mill-bidu tal-Gwerra Dinjija, Gwerra li bidlet lill-Ewropa*, 15 June 2014

**United Kingdom**

Women's Weekly:

*We remember them*, World War I Collector's Issue, 29 July 2014

## PRIVATE COLLECTIONS

Beck family: photos, Beck family tree, Waldemar's letters, recollections

Bonavia family: photos, Charles's letters, recollections

Castaldi family: Anthony's photo, recollections

Gugeri family: photos, Gugeri family history, Waldemar's war diary

# Image Credits

1. Charles Bonavia, Alfred Pisani
2. Waldemar Beck, Beck family
3. Anthony Xuereb, Beck family
4. Waldemar and his daughter Mariella, author's collection
5. Lawrence Beck, Beck family
6. Waldemar as a baby with his mother Carolina, Beck family
7. Waldemar and his mother, Beck family
8. George, Vira and Josie, Beck family
9. Charles as a young boy, Alfred Pisani
10. Carefree days, Beck family
11. Waldemar and Anthony, Beck family
12. The Bonavia family, Alfred Pisani
13. Lea Harding, daughter of Charles's sister, Evelyn Stilon Depiro, Joseph Deguara and Alfred Pisani, son of Charles's sister, Marie Pisani, Alfred Pisani
14. Waldemar's school certificates, Beck family
15. Photo dedicated to Waldemar from his friend A. Griscti, Beck family
16. Waldemar on his graduation, Beck family
17. Passport applications, National Archives, Malta
18. Waldemar before his departure, Beck family
19. A photo dedicated to Waldemar by his uncle, Herman Maempel, Beck family
20. Farewell by family and friends, Beck family
21. Waldemar and Anthony on their way to a brighter future in Australia, Beck family
22. Greetings from Perth, Alfred Pisani
23. Postcard to George and Vira, Beck family
24. Eligio Castaldi, Peter Castaldi
25. New Year wishes from Charles, Alfred Pisani

26. Christmas greetings from Charles to his sisters, Alfred Pisani
27. Velina next to Waldemar before his departure, Beck family
28. Velina, Beck family
29. Postcard from Velina to Waldemar, Beck family
30. Barbara, Barbara Barter
31. Silver locket, Debbie Gugeri
32. Barbara with her mother and sisters, Beck family
33. Charles's enlistment papers, Service Record B2455, Bonavia C.E., National Archives, Canberra
34. *The Daily Malta Chronicle* edition announcing Anthony's enlistment, National Library, Malta
35. Waldemar, Christmas 1914, Barbara Barter
36. Charles in Egypt, Alfred Pisani
37. Charles as Private 157, 11th battalion, John Mizzi, *Gallipoli: the Malta connection*
38. Map of Gallipoli, Alfred Pisani
39. Grave of Giuseppe Camilleri, Maltese Labour Corps, author's collection
40. Anthony's enlistment papers, B2455, Xuereb A., National Archives, Canberra
41. Anthony and his battalion, Northern Territory Library, Territory Stories, PH0142, 0183
42. Anthony forming part of the Cable Guard, Northern Territory Library, Territory Stories, PH0238-1105
43. '*Deceased*' stamp on Charles's record, Service Record B2455, Bonavia C.E., National Archives, Canberra
44. Charles's medals, Noel Zarb Cousin
45. Waldemar's enlistment papers, Service Record B2455, Beck W., National Archives, Canberra
46. Barbara and Waldemar's final moments together, Barbara Barter

47. Barbara, Barbara Barter
48. Waldemar with two of his battalion mates at Blackboy Hill, Barbara Barter
49. Waldemar with his battalion on the day of departure to the Front, Beck family
50. Waldemar, Barbara Barter
51. Josie's endearing postcard to his uncle, Barbara Barter
52. Barbara or Bobbie, Barbara Barter
53. Waldemar with two of his mates in France, Barbara Barter
54. Waldemar's diary, Barbara Barter
55. Extract from Waldemar's diary, Barbara Barter
56. A changed man, Waldemar in December 1916 and April 1918, Barbara Barter
57. Waldemar's service record stamped *'Died of wounds'*, National Archives, Canberra
58. A tribute to Waldemar, *The Swan Express*, 9th August 1918, National Library of Australia (online at www.trove.nla.gov.au)
59. Waldemar's memorial card, Barbara Barter
60. Graves of the fallen, Alfred Pisani
61. George's friend paying his respects at Waldemar's grave, Beck family
62. Anthony's visit to Malta after the war, Peter Castaldi
63. Power of attorney signed by Mr. Bonavia, ITEM 1918/119, State Records Office of Western Australia
64. Waldemar's memorial scroll, Beck family
65. Jack commemorating Charles at Gallipoli, Jack Formosa
66. Peter Castaldi and Charles N. Mifsud, President of the Maltese RSL Sub Branch (NSW), next to the plaque commemorating Maltese Anzacs, Joseph Deguara
67. Anthony's obituary, *Malta News*, 1st May 1948, Mark Caruana
68. Eligio in 1960, Peter Castaldi
69. Barbara in 1950, *The West Australian*, 25th August 1950, National Library of Australia (online at www.trove.nla.gov.au)
70. Carolina, Beck family

The three friends finally at peace:

71. Charles commemorated on Lone Pine Memorial in Gallipoli, Jack Formosa
72. Plaque on Anthony's grave at Toowong cemetery in Brisbane, Brisbane City Cemetries, Brisbane City Council
73. Waldemar's grave at Boisguillaume cemetery, France, the Australian War Graves Photographic Archive – www.australianwargraves.org.

Author's photo on back cover: Darren T. Schembri

ALEXANDRIA
AWARDED TO
BONAVIA
FOUR YE
IN
QUEENS
BY
E. B. KENN
With a Map
A SQUATTERS HOUSE
MALTA
LONDON:
EDWARD STANFORD, 6 & 7, CHARIN
1870.

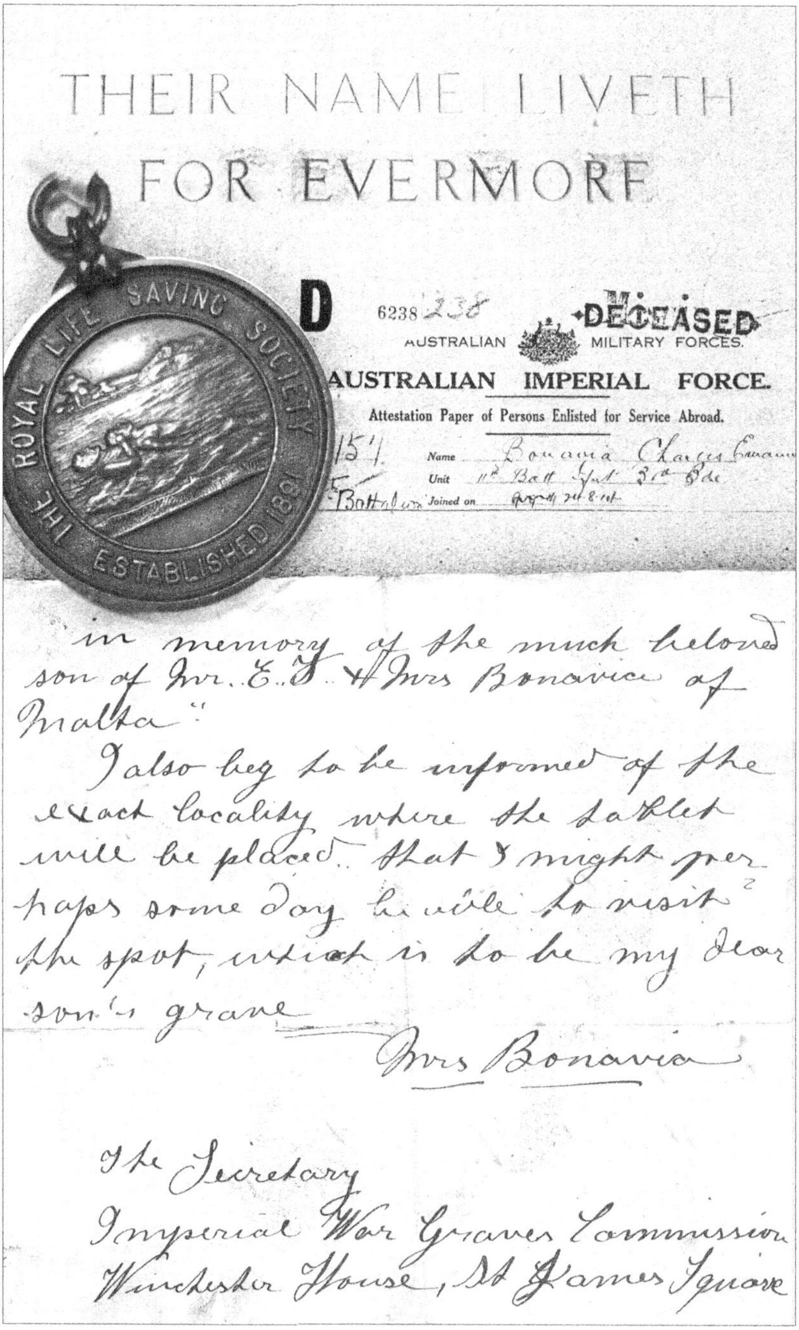


D

6238 238 DECEASED

AUSTRALIAN MILITARY FORCES.

AUSTRALIAN IMPERIAL FORCE.

Attestation Paper of Persons Enlisted for Service Abroad.

151

Name Bonavia Charles Emmanuel

Unit 11th Batt Inf 3rd Bde

Battalion Joined on

"in memory of the much beloved son of Mr. E. J. & Mrs Bonavia of Malta"

I also beg to be informed of the exact locality where the tablet will be placed. that I might perhaps some day be able to visit the spot, which is to be my dear son's grave

Mrs Bonavia

The Secretary
Imperial War Graves Commission
Winchester House, St James Square

ON ACTIVE SERVICE.
191
Y.M.C.A
On Active
George Beck Esq.
The Treasury
Valletta.
WESTERN AUSTRALIA
91 Sda. Ridolfo,
Sliema.
MALTA
No 6969
Pte W BECK
23/16
A.I.F.

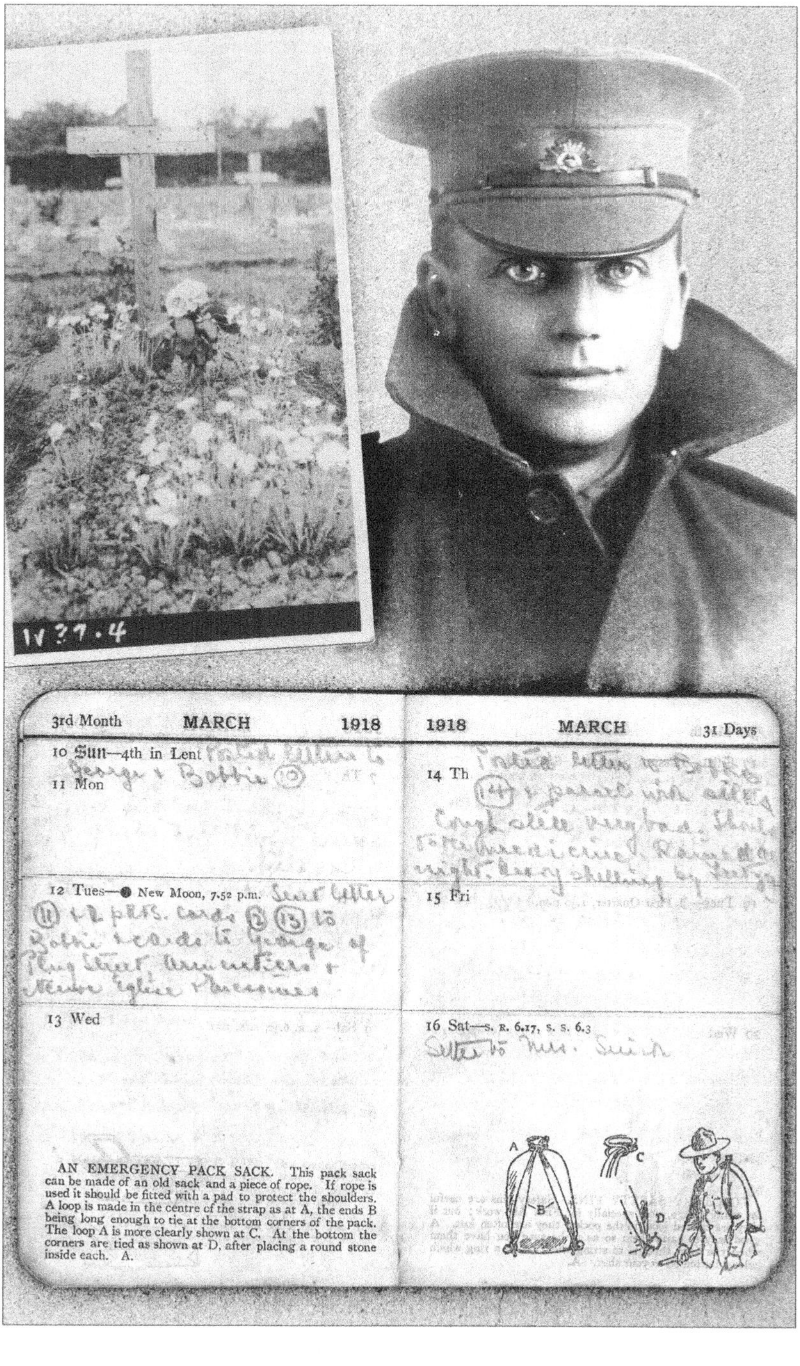
3rd Month MARCH 1918
1918 MARCH 31 Days
10 Sun—4th in Lent Posted letters to George & Bobbie (10)
11 Mon
12 Tues— New Moon, 7.52 p.m. Sent letter (11) + 2 pict. cards (12) (13) to Bobbie & cards to George of Plug Street Armentieres & Neuve Eglise Messines
13 Wed
14 Th
15 Fri
16 Sat—s. r. 6.17, s. s. 6.3
Letter to Mrs. Smith
AN EMERGENCY PACK SACK. This pack sack can be made of an old sack and a piece of rope. If rope is used it should be fitted with a pad to protect the shoulders. A loop is made in the centre of the strap as at A, the ends B being long enough to tie at the bottom corners of the pack. The loop A is more clearly shown at C. At the bottom the corners are tied as shown at D, after placing a round stone inside each. A.
A
B
C
D

CUSTOM HOUSE—MALTA.
Prescribed by Article 183 of the Police Laws.
LIST OF PASSENGERS on board the British Stear "Simla" leaving Malta, for Port Sa
Date
Name and Surname
Nation or Country
Profession Occupation or Trade
Proceeding to
Remarks
Lt John A. C. Legh R.N.
Bombay
Mr Valdemare Beck
Perth W.A.
Mr Ant. Xuereb
Miss Gwendoline Geddes
Aden
And all those on board as per list.
Total 75.
OK
7 Castle S
Fairfield
Brisbane
Qld.
Secretary Department of Army
Canberra A.C.T.
Dear Sir,
I wish to claim my medal and lapel badge for service at Gallipoli. The service detail were
Num. 1055
Pte A. XUEREB
25 Batallion A.I.F.
Yours Sincerely
A Xuereb
PERSONAL
Mr. E. Castaldi, of the firm of Messrs Hickey & Co., Sydney, was in Melbourne on business last month. He was accompanied by Mr. Anthony Xuereb, who has been a member of the Public Service of the Northern Territory since 1914. The two of them were comrades in the first A.I.F. While in Melbourne they met the President (Mr Emanuel Gatt) and other members of the Committee of the Maltese Social Society.

E.G./ G.W.R.
IN REPLY PLEASE QUOTE FILE 18839
INFORMATION BUREAU
FOR OBTAINING NEWS OF SICK, WOUNDED AND MISSING SOLDIERS
AUSTRALIAN RED CROSS SOCIETY
N.S.W. DIVISION
DALTON HOUSE,
115 PITT STREET,
SYDNEY.
25th July, 1917.
XUEREB.
Dear Sir,
We have an enquiry for the whereabouts of a man named Xuereb, a Maltese, who enlisted in Port Darwin in the first year of the war, or early in the second, and who was at Gallipoli.
If with these meagre details, you can supply us with any particulars, we shall be very grateful.
Yours faithfully,
G.W. Rutherford
To The Officer in Charge,
Base Records Office,
Victoria Barracks,
MELBOURNE.
BASE RECORDS
JUL 27 1917
RECEIVED
DEATHS in the District of BRISBANE
in the State of QUEENSLAND
MILITARY FORCES.
A.M. Form D2.
AUSTRALIAN
MEDICAL REPORT ON AN INVALID.
MEDICAL OFFICER'S STATEMENT OF CASE.
One week in Hospital in France Rouen with shell shock
In hospital with sprained ankle in October 1916
Invalid- On disembarkation admitted to Hospital

Made in United States
North Haven, CT
04 November 2026

21760264R00153